SpringerBriefs in Statistics

T0255476

For further volumes:
http://www.springer.com/series/8921

M. B. Rajarshi

Statistical Inference for Discrete Time Stochastic Processes

 Springer

M. B. Rajarshi
Department of Statistics
University of Pune
Pune, Maharashtra
India

ISSN 2191-544X   ISSN 2191-5458   (electronic)
ISBN 978-81-322-0762-7   ISBN 978-81-322-0763-4   (eBook)
DOI 10.1007/978-81-322-0763-4
Springer New Delhi Heidelberg New York Dordrecht London

Library of Congress Control Number: 2012949564

Printed on acid-free paper

Springer is part of Springer Science+Business Media (www.springer.com)

To
Sujata, Girija and Salil

Preface

This book is an overview of statistical inference in stationary, discrete time stochastic processes.

We begin our discussion with martingales and strong mixing sequences. We illustrate how these properties enable us to generate various classes of CAN estimators in the case of dependent observations.

Next, we discuss likelihood inference for finite and infinite Markov chains, higher order Markov chains, Raftery's Mixture Transition Density model and Hidden Markov chains. In Chap. 3, we discuss a number of processes which have a non-Gaussian stationary distribution. Such models can be viewed as extensions of linear Auto-Regressive Moving Average models. Models discussed therein include standard discrete distributions such as Binomial, Poisson, Geometric, and continuous distributions such as Exponential, Gamma, Weibull, Lognormal, Inverse Gaussian and Cauchy.

Chapter 4 deals with semi-parametric methods of estimation wherein few conditional moments are specified and the form of underlying distribution is not specified. Here, Conditional Least Squares methods are discussed. The main theme of the chapter is the estimation and confidence interval procedures based on estimating functions.

In the last two chapters, we discuss non-parametric methods of estimation. In Chap. 5, kernel-based estimation of density and conditional expectation are discussed. Here, it is assumed that the underlying process is strong mixing. Asymptotic normality of these estimators is reported therein. The last chapter has a discussion on bootstrap and other resampling procedures for dependent sequences such as Markov chains, Markov sequences, linear Auto-Regressive Moving Average sequences. Block-based bootstrap for stationary sequences and other block-based procedures are discussed in some details. The main result reported therein is that block-based bootstrap, under certain conditions, is a better approximation to the sampling distribution than the traditional normal approximation. The discussion is concluded by bootstrap procedures for confidence intervals based on estimation functions.

This book can be useful for researchers interested in knowing developments in inference in discrete time stochastic processes. It can be used as a material for advanced level research students. A good background of probability, asymptotic inference and stochastic processes is desirable.

Acknowledgments

Thanks are due to Dr. N. Balakrishana (CUSAT, Cochi, India), Dr. U. V. Naik-Nimbalkar and Dr. T. V. Ramanathan (both from University of Pune, India) for a careful reading of the draft of the book and constructive criticism which led to a considerable improvement.

I would like to thank Springer and all those associated with the Briefs series, for their kind co-operation throughout the writing of the book.

Lastly, I sincerely thank Miss Akanksha Kashikar and Mrs. Inderdeep Kaur for reading several versions of the book and for making a number of useful suggestions. They also skilfully (and patiently) dealt with my adventures in LaTeX.

Pune, India, May 2012   M. B. Rajarshi

Contents

Acronyms

ACF	Auto-Correlation Function
AR	Auto-Regressive
ARMA	Auto-Regressive Moving Average
a.s.	Almost surely
BLUE	Best Linear Unbiased Estimator
CAN	Consistent and Asymptotically Normal
c.i.	Confidence interval
CLS	Conditional Least Squares
CLSE	Conditional Least Squares Estimator
Cov	Covariance
DARMA	Discrete Auto-Regressive Moving Average
d.f.	Distribution function
EF	Estimating Function
FMSE	Forecasting Mean Squared Error
GSM	Geometrically Strong Mixing
i.i.d.	Independently and identically distributed
INAR	Integer-valued Auto-regressive
LRT	Likelihood Ratio Test
LS	Least Squares
MA	Moving Average
ML	Maximum Likelihood
MLE	Maximum Likelihood Estimator
MSE	Mean Squared Error
OLS	Ordinary Least Squares
PACF	Partial Auto-Correlation Function
p.d.f.	Probability density function
RMSE	Root Mean Squared Error
SRSWR	Simple Random Sampling With Replacement
t.p.m.	Transition probability matrix
Var	Variance
WLS	Weighted Least Square

Chapter 1
CAN Estimators from Dependent Observations

Abstract In this chapter, we review some basic properties of stationary stochastic processes. Results on martingale limit theorems and laws of large numbers for mixing sequences, as well as central limit theorems for sums of dependent random variables have been discussed. We then discuss weak convergence of empirical processes obtained from stationary observations. These results have been applied to generate consistent and asymptotically normal estimators of parameters of a stationary stochastic process.

1.1 Preliminaries

A stochastic process is a collection of random variables $\{X_t, t \in \mathcal{T}\}$ on a probability space $\{\Omega, \mathscr{A}, P\}$. The set $\mathcal{T}$ is an infinite set, known as time parameter set. Let S_t be the collection of values taken by X_t. The set $S = \cup_t S_t$ is known as the state-space of $\{X_t, t \in \mathcal{T}\}$. Throughout our discussion, S is either $\Re^p$, the p-dimensional Euclidean space or a countable set. The set $\mathcal{T}$ is assumed to be a countable set which is frequently $\{0, 1, 2, \ldots\}$ or its subset. Thus, we have a discrete time stochastic process or a random sequence to be denoted by $\mathfrak{X}$.

Definition 1.1.1 A stochastic process $\mathfrak{X}$ is said to be strictly stationary, if the distribution of $(X_{t_1}, \ldots, X_{t_n})$ is the same as that of $(X_{t_1+h}, \ldots, X_{t_n+h})$ $\forall\, t_1 < t_2 < \cdots < t_n$, $\forall\, h$ and $\forall\, n$.

The symbol V denotes variance of a random variable or variance-covariance matrix of a random vector. The symbol Cov denotes covariance between two random variables.

Definition 1.1.2 A stochastic process $\mathfrak{X}$ is said to be weakly stationary, if $V(X_t) < \infty$ for all t and the followings hold.

1. $E(X_t) = \mu$ and $V(X_t) = \sigma^2$, $\forall\, t$.

M. B. Rajarshi, *Statistical Inference for Discrete Time Stochastic Processes*,
SpringerBriefs in Statistics, DOI: 10.1007/978-81-322-0763-4_1,
© The Author(s) 2012

2. $Cov(X_s, X_t)$ depends on s and t through $| s - t |$ only.

Definition 1.1.3 A random sequence $\mathcal{X}$ is said to be a first order Markov sequence if

$$P[X_t \in B \mid X_{t-1} = x_{t-1}, X_{t-2} = x_{t-2}, \dots, X_0 = x_0] = P[X_t \in B \mid X_{t-1} = x_{t-1}],$$

for all $t \geq 1$, for all Borel sets B of S and for all $\{x_{t-1}, x_{t-2}, \dots, x_0\}$ such that the conditional probability on the left-hand side is well-defined.

Higher order Markov sequences can be similarly defined.
A Markov sequence $\mathcal{X}$ is said to be time-homogeneous if for all s, t such that $t > s$, $P[X_t \in B \mid X_s = x_s]$ is a function of $t - s$ only. If S is countable, $\mathcal{X}$ is called a Markov chain and we have $P = ((p_{ij}))$ as the one-step transition probability matrix (t.p.m.), where $p_{ij} = P[X_{t+1} = j | X_t = i]$. If S is $\Re^p$, and if the distribution of X_{t+1} given X_t is absolutely continuous, the corresponding density $f(y|x)$ is known as the (one-step) transition density.

Definition 1.1.4 A probability mass function (p.m.f.) π on a countable set S is known as a stationary distribution of a time-homogeneous Markov chain $\{X_t, t \geq 0\}$, if

$$\pi_j = \sum_{i \in S} \pi_i p_{ij}, \quad \text{for all } j \in S. \tag{1.1}$$

A probability density function (p.d.f.) π on $S \subseteq \Re^p$, is said to be a stationary density of a time-homogeneous Markov sequence $\{X_t, t \geq 0\}$ with the transition density $f(y|x)$, if

$$\pi(y) = \int_{x \in S} \pi(x) f(y|x) dx, \quad \text{for all } y \in S.$$

The following theorems discuss application of the ergodic theorem to obtain almost sure convergence of averages of functions of observations from a Markov chain.

Theorem 1.1.1 *If $\{X_t, t = 0, 1, 2, \dots\}$ is an irreducible, time-homogeneous and non-null persistent Markov chain with $\{\pi_i, i \in S\}$ as the stationary distribution, then $\mathcal{X}$ is strictly stationary if the distribution of X_0 is given by $\{\pi_i, i \in S\}$.*

For a background of ergodic theorems, we refer to Sect. 8.6 of Athreya and Lahiri (2006).

Theorem 1.1.2 *Let $\{X_t, t = 0, 1, 2, \dots\}$ be an irreducible, aperiodic, and non-null persistent stationary Markov chain and let g be a function on S such that $E(|g(X_0)|) < \infty$. Then $\frac{1}{T+1} \sum_{t=0}^{T} g(X_t) \to E(g(X_0))$ a.s.*

If a Markov sequence $\mathcal{X}$ is metrically transitive, i.e., the conditional distribution (probability measure corresponding to the transition density $f(y|x)$) is absolutely continuous with respect to the stationary distribution (for every $x \in S$), then almost sure convergence as in Theorem 1.1.2 holds [cf. Chap. 9 of Billingsley (1961a)].

1.2 Martingales

Definition 1.2.1 *Martingale.* Let $\{X_t, t \geq 0\}$ be a sequence of random variables defined on a probability space $\{\Omega, \mathscr{A}, P\}$. Let $\{\mathscr{F}_t, t \geq 0\}$ be a family of non-decreasing sub-σ-fields, i.e., $\mathscr{F}_t \subset \mathscr{F}_{t+1} \subset \mathscr{A}$ $\forall\, t$. Then, $\{X_t, t \geq 0\}$ is said to be a martingale with respect $\{\mathscr{F}_t, t \geq 0\}$, if for all positive t and s, we have

$$E(X_{t+s} \mid \mathscr{F}_s) = X_s \ a.s.$$

Frequently, $\mathscr{F}_s = \sigma\{Y_0, \ldots, Y_s\}$, where $\{Y_t, t \geq 0\}$ is a stochastic process defined on $\{\Omega, \mathscr{A}, P\}$. We give now some examples of martingale sequences.

(i) *Symmetric Random Walk.* Let Y_t be a sequence of i.i.d. random variables with $E(Y_t) = 0$. Then, $X_T = \sum_{t=1}^{T} Y_t$ is a martingale.

(ii) Suppose we have a family of parametric models $\{P_\theta, \theta \in \Theta\}$ for the sequence $\mathfrak{X}$, where Θ is an open interval of $\mathfrak{R}$. We assume that the conditional p.d.f./p.m.f. $f_\theta(x_t \mid x_0, \ldots, x_{t-1})$ satisfies the condition that $\int f_\theta(u \mid x_0, \ldots, x_{t-1}) du$ can be differentiated twice with respect to θ under the integral sign and that $E[\frac{\partial \log f_\theta(X_t \mid X_0, \ldots, X_{t-1})}{\partial \theta}]^2 < \infty$ for all t. Let

$$S_t = \frac{\partial \log f_\theta(X_t \mid X_0, \ldots, X_{t-1})}{\partial \theta}$$

and $Z_t = \sum_{s=1}^{t} Z_s$. Then, $E[S_t \mid \sigma\{X_0, \ldots, X_{t-1}\}] = 0$ and $\{Z_t, t \geq 1\}$ is a martingale with respect to family of σ-fields of $(X_0, X_1, \ldots, X_t)$, $t \geq 0$.

(iii) Let $f_\theta(x_0, x_1, \ldots, x_t)$ be the sequence of p.d.f.s or p.m.f.s of $(X_0, \ldots, X_t)$, when θ is the parameter. Suppose that $H_0 : \theta = \theta_0$ and $H_1 : \theta = \theta_1$ are the null and the alternative hypotheses respectively. Let $L_t = \frac{f_{\theta_1}(X_0, X_1, \ldots, X_t)}{f_{\theta_0}(X_0, X_1, \ldots, X_t)}$. Then, $\{L_t, t \geq 1\}$ is a martingale under H_0.

Theorem 1.2.1 *Martingale Convergence Theorem. Suppose that $\{X_t, t \geq 1\}$ is a uniformly integrable martingale. Then, there exists a random variable X such that $X_t \to X$ a.s.*

For a proof of the above theorem, we refer to Athreya and Lahiri (2006), Chap. 13. We note that if for some $p > 1$, $E(|X_t|)^p < C < \infty$ for all t, $\{X_t, t \geq 1\}$ is uniformly integrable. The following theorem states a central limit theorem (CLT) for martingales.

Theorem 1.2.2 *Billingsley-Ibragimov CLT for Stationary Martingales Billingsley (1961a) and Ibragimov (1963). Let $\{Y_t, t \geq 0\}$ be a strictly stationary ergodic sequence. Let $X_T = \sum_{t=1}^{T} Y_t$ and assume that $E[Y_{t+1} \mid X_0, \ldots, X_t] = 0$ for all $t \geq 1$ and that $E(Y_1) = 0$. Let $\sigma^2 = Var(Y_t) < \infty$. Then,*

$$\frac{X_T}{\sqrt{T}} \overset{\mathscr{D}}{\to} N(0, \sigma^2).$$

Example 1.2.1 Let an auto-regressive sequence $\mathfrak{X}$ be defined by $X_{t+1} = \rho X_t + \varepsilon_{t+1}$, $|\rho| < 1$, where $\{\varepsilon_t\}$ is an i.i.d. sequence which is also independently distributed of X_0. Let $E(\varepsilon_t) = 0$, $V(\varepsilon_t) = \sigma^2$. Then, $E[X_{t+1}|X_0, X_1, \ldots, X_t] = \rho X_t$ and thus, $\{Y_t, t \geq 1\}$ defined by $Y_t = (X_t - \rho X_{t-1})X_{t-1}$ is a martingale. If X_0 follows the stationary distribution, i.e., the distribution of $\sum \rho^t \varepsilon_t$, the sequence $\mathfrak{X}$ is strictly stationary. It can also be shown to be weakly stationary. Thus,

$$\frac{\sum_{t=1}^{T} Y_t}{\sqrt{T}} \xrightarrow{\mathscr{L}} N\left(0, \sigma^2 E(X_1^2)\right).$$

Let $\hat{\rho} = \sum X_t X_{t-1}/\sum X_{t-1}^2$. By the ergodic theorem, $\sum_{t=1}^{T} Y_t/T \to 0$ a.s. and it follows that $\hat{\rho} \to \rho$ a.s. Further, by the ergodic theorem, $\frac{\sum_{t=1}^{T} X_t^2}{T} \to E(X_1^2) = \frac{\sigma^2}{1-\rho^2}$ a.s. By the Martingale CLT (Theorem 1.2.2), we have

$$\frac{\sum[(X_t - \rho X_{t-1})X_{t-1}]}{\sqrt{T}} \xrightarrow{\mathscr{L}} N\left(0, \frac{\sigma^4}{1-\rho^2}\right).$$

Consequently,

$$\sqrt{T}(\hat{\rho} - \rho) \xrightarrow{\mathscr{L}} N(0, 1-\rho^2). \tag{1.2}$$

Martingale limit theorems, CLTs for martingales, and statistical applications of martingale asymptotics have been discussed in Hall and Heyde (1980). It is convenient to have the following definition.

Definition 1.2.2 $\sqrt{T}$-*Consistent and Asymptotically Normal* A sequence of estimators $\hat{\theta}_T$ of θ is said to be $\sqrt{T}$-CAN if, as $T \to \infty$, $\hat{\theta}_T \to \theta$ in probability/ a.s. and the distribution of $\sqrt{T}(\hat{\theta}_T - \theta)$ weakly converges to a normal distribution with mean 0.

1.3 Mixing Sequences

In statistical analysis of stationary sequences, we have information on the conditional distribution or moments of an observation given the past observations or on the marginal distribution of observation only. Thus, martingale strong laws of large numbers and martingale central limit theorems play a prominent role, as seen in Example 1.2.1. If our information (or assumption) of a statistical model consists of marginal distributions (or functions thereof) only, mixing properties play a major role. These are discussed below.

Let $\{\Omega, \mathscr{A}, P\}$ be a probability space and let $\mathscr{F}$ and $\mathscr{G}$ be sub-σ-fields of $\mathscr{A}$. Various dependence coefficients between $\mathscr{F}$ and $\mathscr{G}$ are defined below.

$$\alpha(\mathcal{F}, \mathcal{G}) = \sup_{\substack{A \in \mathcal{F} \\ B \in \mathcal{G}}} P(A \cap B) - P(A)P(B) \mid .$$

$$\beta(\mathcal{F}, \mathcal{G}) = E\left[\sup_{B \in \mathcal{G}} \mid P(B \mid A) - P(B) \mid \right], \quad A \in \mathcal{F}.$$

$$\phi(\mathcal{F}, \mathcal{G}) = \sup_{\substack{A \in \mathcal{F}, \ B \in \mathcal{G} \\ P(A) > 0}} \mid P(B \mid A) - P(B) \mid .$$

Let $L^2(\mathcal{H})$ be the collection of $\mathcal{H}$-measurable square integrable random variables. Let $Cor(X, Y)$ denote the correlation co-efficient between the random variables X and Y. Then,

$$\rho(\mathcal{F}, \mathcal{G}) = \sup_{\substack{X \in L^2(\mathcal{F}) \\ Y \in L^2(\mathcal{G})}} Cor(X, Y).$$

The α-mixing coefficient was introduced by Rosenblatt, the β-mixing coefficient was introduced by Kolmogorov, the ϕ-mixing coefficient was introduced by Ibragimov and the ρ-mixing coefficient was introduced by Kolmogorov. For these and related earlier references as well as an extensive account of mixing sequences, we refer to Doukhan (1994) and Bradley (2005).

Let $\{X_t, t = 0, \pm 1, \pm 2, \ldots\}$ be a strictly stationary sequence. Let $\sigma\{X, Y, Z, \ldots\}$ denote the σ-field generated by the collection of random variables $\{X, Y, Z, \ldots\}$.

$$\alpha(t) = \alpha\big(\sigma\{X_s, X_{s-1}, X_{s-2}, \ldots\}, \sigma\{X_{s+t}, X_{s+t+1}, \ldots\}\big), \quad \forall s.$$

Definition 1.3.1 A stationary sequence $\{X_t, t = 0, \pm 1, \pm 2, \ldots\}$ is said to be strong mixing or α-mixing, if $\alpha(t) \to 0$ as $t \to \infty$.

Mixing sequences such as β-mixing, ϕ-mixing and ρ-mixing can be similarly defined. In our discussion, we assume that the given stationary sequence is α-mixing, since each of the β-mixing, ϕ-mixing and ρ-mixing sequences imply that the sequence is α-mixing, cf. Doukhan (1994). Secondly, α-mixing sequences have been widely discussed in the literature and a large number of results are available for such sequences.

Results on estimation of density and estimation of conditional expectation (also known as regression function) as well as on bootstrap have been recently proved based on some newly defined types of weak dependence, cf. Bickel and Bühlmann (1999), Nze, Bühlmann and Doukhan (2002) and Nze and Doukhan (2002). For a detailed review of mixing sequences, we refer to Bradley (2005).

In the case of independent observations, $\alpha(t) = 0$ for all $t \geq 1$. We follow Bosq (1996) and define a geometrically strong mixing as follows.

Definition 1.3.2 We say that a stationary process $\mathfrak{X}$ is Geometrically Strong Mixing (GSM) if for some constant C and β, $0 < \beta < 1$, the process is α-mixing with $\alpha(t) \leq C\beta^t$, $t \geq 1$.

Examples of mixing sequences

Example 1.3.1 An *m*-dependent sequence

A stationary sequence $\mathfrak{X}$ is said to be an *m*-dependent sequence, if the collection of random variables $\{X_s, s \leq t\}$ is independent of the collection $\{X_s, s \geq t + m\}$ for every t. Such a sequence is α-mixing with $\alpha(t) = 0$ for all $t > m$. For example, a moving average sequence defined by $X_t = \sum_{s=1}^{q} a_s \varepsilon_{t-s}$, where $\{\varepsilon_t\}$ is a sequence of independently and identically distributed (i.i.d) random variables with mean 0 and $E[\varepsilon_t \varepsilon_s] = 0$, $s \neq t$,is a *q*-dependent sequence.

Example 1.3.2 A finite irreducible and aperiodic Markov chain

A finite, irreducible, and aperiodic Markov chain is ϕ-mixing with a geometric decay of the ϕ-mixing coefficients, cf. page 166 of Billingsley (1968). Hence, such a process is α-mixing and in fact, GSM . It follows that an *r*-order finite irreducible and aperiodic Markov chain is GSM, though the first *r* mixing coefficients may not have such a geometric rate.

Example 1.3.3 Markov sequences

Suppose that the transition density $f(y|x)$ of a Markov sequence satisfies the condition that

$$\sup_{x,x' \in S, A \subset S} \left| \int_A f(y|x)dy - \int_A f(y|x')dy \right| < 1.$$

It is assumed that A is measurable. Then, the Markov sequence is GSM (cf. Götze and Hipp (1983)). The above condition is satisfied, if there exists a positive measure μ on S such that $\int_A f(y|x)dy \geq \mu(A)$ for all x and A.

Example 1.3.4 A counter-example

Let $\{\varepsilon_t, t = 0, \pm 1, \pm 2, \ldots\}$ be an i.i.d. sequence of Bernoulli random variables with $P[\varepsilon_t = 1] = 1/2$. Define the sequence $\{X_t\}$ by

$$X_t = \sum_{s=0}^{\infty} 2^{-s-1} \varepsilon_{t-s}.$$

It can be shown that the sequence X_t is strictly stationary with $U(0, 1)$ as the stationary distribution. But $2X_{t+1} = X_t + \varepsilon_{t+1}$ and X_t is the fractional part of $2X_{t+1}$. Thus, X_t can be uniquely computed ("recovered"), if we know the future. Thus $\sigma\{X_t\}$ is included in $\sigma\{X_{t+1}\}$ and $\alpha_1 \geq 1/4$, cf. Bosq (1996), page 16. Since $\alpha(t) \leq 1/4$ for all t, it follows that the process $\{X_t\}$ is not mixing. A similar result holds for any value of $p = P[\varepsilon_t = 1]$.

Example 1.3.5 General linear sequences

We now discuss results due to Gorodetskii (1977). Let $\{\varepsilon_t, t = 0, \pm 1, \pm 2, \ldots\}$ be i.i.d. random variables with p.d.f. $f(x)$. The process $\{X_t, t = 0, \pm 1, \pm 2, \ldots\}$ is defined by

$$X_t = \sum_{s=0}^{\infty} a_s \varepsilon_{t-s}.$$

Let

$$A_t(\delta) = \sum_{s=t}^{\infty} |a_s|^\delta,$$

$$B_t = \begin{cases} \displaystyle\sum_{s=t}^{\infty} [A_s(\delta)]^{1/(1+\delta)}, & \delta < 2 \\ \displaystyle\sum_{s=t}^{\infty} \max\left\{[A_s(\delta)]^{1/(1+\delta)}, \sqrt{A_s(2) |\ln(A_s(2))|}\right\}, & \delta \geq 2. \end{cases}$$

Theorem 1.3.1 *(Gorodetskii 1977). Assume the following.*

1. $\displaystyle\int_{-\infty}^{\infty} |f(x+y) - f(x)| \, dx \leq C |y|$.

2. $E(|\varepsilon_1|^\delta) \leq C < \infty$ *for some* $\delta > 0$. *If* $\delta \geq 1$ *we assume that* $E(\varepsilon_1) = 0$. *For* $\delta \geq 2$, *we assume that* $Var(\varepsilon_1) = 1$.

3. *Let* $g(z) = \sum_{t=0}^{\infty} a_t z^t$. *Then,* $g(z)$ *does not vanish in* $|z| \leq 1$.

4. $B_0 < \infty$.

Then, $\{X_t, t = 0, \pm 1, \pm 2, \ldots\}$ *is* α-*mixing with* $\alpha(t) \leq M B_t$, *for some real positive* M.

We notice that if the theorem holds with the restriction that $\delta < 1$, such a process can not be described as second order or weakly stationary.
Let us assume that $Var(\varepsilon_t) = \sigma_\varepsilon^2 < \infty$, $\sum a_t \neq 0$ and $a_s = O(e^{-\gamma s})$, $\gamma > 0$. Further, assume that ε_t has an absolutely continuous distribution. It has been shown that $\{X_t, t = 0, \pm 1, \pm 2, \ldots\}$ is β-mixing and in fact, GSM.
Such processes include ARMA(p, q), AR(p) and MA$((q))$ processes, if all the roots of each of the related AR and MA polynomials lie outside the unit circle cf. Withers (1981) and Athreya and Pantula (1986b).

Example 1.3.6 Harris-recurrent Markov sequences

Let $\mathcal{X}$ be a Markov sequence with the state-space S. A Markov chain is said to be Harris-recurrent, if there exists a non-trivial σ-finite measure μ on S such that $\mu(A) > 0$ implies that $P[X_t \in A$ for some $t \geq 1 | X_0 = x] = 1$ for every $x \in S$. Suppose that the Markov sequence has a unique stationary distribution. Athreya and Pantula (1986a) have shown that such a sequence is strong mixing.

Example 1.3.7 Nonlinear time series

These models have been discussed in Tjøstheim (1994). Let $\{\varepsilon_t\}$ be a sequence of i.i.d. random variables with mean 0 and variance 1. Define

$$X_t = M(X_{t-1}, X_{t-2}, \ldots, X_{t-p}, \theta_1) + V(X_{t-1}, X_{t-2}, \ldots, X_{t-p}, \theta_2)\varepsilon_t.$$

The functions $M(X_{t-1}, X_{t-2}, \ldots, X_{t-p}, \theta_1)$ and $V^2(X_{t-1}, X_{t-2}, \ldots, X_{t-p}, \theta_2)$ are respectively the conditional mean and conditional variance of X_t given the past. It is assumed that $V > 0$ a.s. for all $(X_{t-1}, X_{t-2}, \ldots, X_{t-p})$. Here θ_1, θ_2 are vector parameters.

Some examples of non-linear models of order 1 are given below.

1. $M(x, \theta) = \theta_1 x I(x \leq k) + \theta_2 x I[x > k]$ (threshold model)
2. $M(x, \theta) = x\{\theta_1 + \theta_2 \exp(-\theta_3 x^2)\}, \quad \theta_3 > 0$ (exponential auto-regression model)
3. $M(x, \theta) = \theta_1 x + \theta_2 x\{[1 + \exp\{-\theta_3(x - \theta_4)\}]^{-1} - 1/2\}, \quad \theta_3 > 0$ (logistic AR model).

These models are very flexible and exhibit rich patterns of stochastic behaviour.

Here, we focus on the case $V^2 = \sigma^2$, a constant. We assume that

1. The function M is bounded on compact sets.
2. The function M satisfies the condition that $M(x) = a'x + o(\| x \|)$ as $\| x \| \to \infty$ and the linear model $a'x = \sum_{i=1}^{p} a_i x_i$ is stable in the sense $z^p - \sum_{i=1}^{p} a_i z^{p-i}$ has its zeroes in the unit circle (it is possible that the vector a is null).
3. The p.d.f. of ε_1 is positive on $\Re$ and $E(| \varepsilon_1 |) < \infty$.

Then, the Markov sequence $\mathfrak{X}$ is geometrically ergodic, so that the t-step transition probability converges to the invariant distribution at the geometric rate as $t \to \infty$. Consequently, $\mathfrak{X}$ is strong mixing. Tjøstheim (1994) discusses a number of interesting examples of nonlinear functions M.

More examples of strong mixing sequences are discussed in the next chapter. We state below some important theorems for strong mixing sequences.

Theorem 1.3.2 (*Davydov's inequality*) *Let Y be a $\sigma\{X_0, X_1, \ldots, X_s\}$-measurable random variable and let Z be a $\sigma\{X_{s+t}, X_{s+t+1}, \ldots\}$-measurable random variable. Suppose that $E \mid Y \mid^p < \infty$ and $E \mid Z \mid^q < \infty$ for some p and q such that $1/p + 1/q < 1$. Let $r = (1/p + 1/q)^{-1}$. Then, $E(YZ) < \infty$ and*

$$\mid Cov(Y, Z) \mid \leq 2r[2\alpha(t)]^{1/r}(E \mid Y \mid^p)^{1/p}(E \mid Z \mid^q)^{1/q}.$$

The following result establishes that under appropriate moment conditions, limit of variance of a mean-like statistic is finite.

Theorem 1.3.3 *Suppose that* $\{X_t, t = 0, 1, 2, \ldots\}$ *is* α-*mixing such that for a* $\delta > 0$,

$$E \mid X_t \mid^{2+\delta} < \infty \quad and \quad \sum_{t=1}^{\infty} \alpha(t)^{\delta/(2+\delta)} < \infty.$$

Then, the series $\sum_{t=2}^{\infty} Cov(X_1, X_t)$ *converges absolutely.*

We observe that, in view of the second order stationarity of the process $\{X_t\}$,

$$Var(\bar{X}_T) = \frac{1}{T} Var(X_1) + \frac{2}{T} \sum_{t=2}^{T-1} (T - t) Cov(X_1, X_t).$$

Now, $\qquad \frac{2}{T} \left| \sum_{t=2}^{T-1} (T - t) Cov(X_1, X_t) \right| < 2r[2\alpha(t)]^{1/r} \left(E \mid X_t \mid^{2+\delta} \right)^{1-1/r}$,

in view of the above inequality with $r = (2+\delta)/\delta$. Thus, $Var(\bar{X}_T) \to 0$ as $T \to \infty$. We further note that as $T \to \infty$,

$$Var(\sqrt{T}\bar{X}_T) \to Var(X_1) + 2 \sum_{t=2}^{\infty} Cov(X_1, X_t) = \sigma^2 \text{ (say)}.$$

If there exists a C such that $\mid X_t \mid < C$, the condition on the α-mixing coefficients can be relaxed to $\sum_{t=1}^{\infty} \alpha(t) < \infty$. We now state a CLT for the sample mean.

Theorem 1.3.4 *(CLT for the sample mean obtained from a stationary* α-*mixing sequences) Suppose that, in addition to the assumptions of Theorem 1.3.3,* $0 < \sigma^2 < \infty$. *Then,*

$$\sqrt{T}(\bar{X}_T - \mu) \xrightarrow{\mathscr{D}} N(0, \sigma^2).$$

Proofs of the Theorems 1.3.2, 1.3.3, and 1.3.4 have been given in Doukhan (1994), also see Athreya and Lahiri (2006), Chap. 16.
Combining the above two theorems, it follows that $\bar{X}_T \to \mu$, in quadratic mean and that $\bar{X}_T$ is a $\sqrt{T}$-CAN estimator of μ.

1.4 Empirical Processes of Dependent Observations

Let $\mathfrak{X}$ be a stationary random sequence with the state-space $\mathfrak{R}^p$. Let $(X_1, X_2, \ldots, X_T)$ be the observations from such a process. Let $x = (x_1, x_2, \ldots, x_p)$ and $y = (y_1, y_2, \ldots, y_p)$. We say that $x \leq y$ if and only if $x_i \leq y_i$ for each i. Let the common distribution function F and the empirical distribution function F_T be respectively defined by

$$F(x) = P[X_1 \leq x]; \quad F_T(x) = \frac{1}{T} \sum_{t=1}^{T} I[X_t \leq x].$$

The empirical process is defined by

$$G_T(x) = \sqrt{T}[F_T(x) - F(x)], \quad x \in \Re^p.$$

Let $\mathscr{D}$ be the space of real valued functions on $\Re^p$, which are continuous from above and for which the limit from below exists at each point. Let $\mathscr{D}$ be equipped by the Skorohod metric. Weak convergence of stochastic sequences taking values in $\mathscr{D}$ has been discussed in Billingsley (1968).

Theorem 1.4.1 *Bühlmann (1994) Assume that the stationary process $\mathfrak{X}$ with the state-space $\Re^p$ is α-mixing with*

$$\sum_{t=1}^{\infty} t^{8p+7} \alpha(t)^{1/2} < \infty.$$

Further, assume that the distribution of X_1 is continuous. Then, $G_T \to G$ weakly, where G is a mean zero Gaussian process with a.s. continuous sample paths such that $E(G(x)G(y)) = \sum_{t=-\infty}^{\infty} Cov(I[X_1 \leq x], I[X_{t+1} \leq y])$.

Earlier results in this direction are due to Yoshihara (1975). Deo (1973) proved the weak convergence of empirical processes obtained from real valued strong mixing sequences and Billingsley (1968), for ϕ-mixing sequences. Withers (1975) proves such results for possibly non-stationary sequences which satisfy various mixing types. We also refer to Radulović (2002) for a discussion of aspects of weak convergence of empirical processes.

In non-parametric analysis, weak convergence of empirical processes plays an important role. Empirical processes are themselves of interest in some applications. Moreover, a large number of statistics can be written as suitable functions of the empirical distribution function and the above weak convergence result allows a derivation of their asymptotic distribution, as is discussed below.

Suppose that θ, a parameter of interest can be written as $H(F)$. Its natural estimator is then $\hat{\theta} = H(F_T)$. A smoothness requirement of θ is described by its differentiability with respect to F in the following manner.

Definition 1.4.1 Let H be a functional defined on the space $\mathscr{F}$ of distribution functions in $\Re^p$, taking values in $\Re^k$. Let Δ denote the space $\{F - G \mid F, G \in \mathscr{F}\}$ and let η be a norm on Δ. Let $\| x \|$ be the usual Euclidean metric on $\Re^p$. Then, the functional H is said to be Fréchet differentiable at $F \in \mathscr{F}$, if there exists a function $h(F, \cdot) : \Delta \to \Re^k$ satisfying the following.

(i) $h(F, \cdot)$ is linear in the sense that

$$h(F, a\gamma_1 + b\gamma_2) = ah(F, \gamma_1) + bh(F, \gamma_2)$$

for all real a, b and for all $\gamma_1, \gamma_2 \in \Delta$.
(ii) Let $G \in \mathscr{F}$. Then,

$$\frac{\| H(F) - H(G) - h(F, G - F) \|}{\eta(F - G)} \to 0,$$

as $\eta(F - G) \to 0$.

The function h is said to be the *Fréchet differential of H at F*.

In applications, $\eta(F - G) = \sup_{x \in \mathfrak{R}} | F(x) - G(x) |$, the Kolmogorov distance between the two distribution functions and $G = F_T$. When weak convergence of empirical process holds, $\eta(G - F_T)$ converges to 0, in probability. Let $U(X_t)$ be the d.f. of the random vector X_t and $V_t = h(F, U(X_t) - F)$. We then have the following.

Theorem 1.4.2 *Suppose that H is Fréchet differentiable with h as the Fréchet differential at F. Assume that $E(V_1) = 0$, $E \parallel V_1 \parallel^3 < \infty$. Let $\Sigma = \lim_{T \to \infty} Var \left[T^{-1/2} \sum_{t=1}^{T} V_t \right]$. Assume that Σ is non-singular. Further, we assume that conditions of Theorem 1.4.1 hold. Then,*

$$\sqrt{T}(\hat{\theta}_T - \theta) \xrightarrow{\mathscr{L}} N_k(0, \Sigma).$$

For a proof of the above results, we refer to Lahiri (2003), Sect. 4.4.2.
Now we discuss the asymptotic normality of L estimators . Suppose that X_t is a real-valued random variable with the d.f. F. We are interested in estimating the parameter $\theta = \int J(u)F^{-1}(u)$. Its estimator is of the type $\hat{\theta} = \int J(u)F_T^{-1}(u)$ and is known as an L estimator. Let F^{-1} be non-decreasing and left continuous (so that F^{-1} induces a measure on $\mathfrak{R}$). We assume that the function $J : (0, 1) \to \mathfrak{R}$ satisfies the following conditions.

1. The function J is bounded and continuous almost everywhere with respect to F^{-1} and Lebesgue measure.
2. There exist $0 < a < b < 1$ such that $J(u) = 0$ for all $u \notin [a, b]$.

It has been shown by Boos (1979) that the above functional is Fréchet differentiable at F with respect to the sup-norm. An important example is the (100α) per cent trimmed mean defined by $\theta = \int_{\alpha}^{1-\alpha} F^{-1}(u)du/(1 - 2\alpha)$ where $0 < \alpha < 0.5$. It may be noted that, if F is symmetric, for any $\alpha \in (0, 0.5)$, the parameter θ turns out to be the median of X_1, cf. Serfling (1980), page 237.

Hadmard differentiability and sample median
It is to be noted that the median of X_t cannot be regarded as an L estimator. It can be shown that median is a Hadamard differentiable statistical functional. The asymptotic normality of Hadamard differentiable statistical functionals (and thus that of the sample median) follows from the weak convergence of the empirical processes, see Fernholz (1983). For a discussion of statistical differentiable functionals, we refer to Sect. 5.2 of Shao (1999).

1.5 CAN Estimation Under Cramér and Other Regularity Conditions

In this section, we prove a general result concerning estimators obtained from $X_0, X_1, \ldots, X_T$ where $\{X_t, t \geq 0\}$ is a discrete time strictly stationary stochastic process. Let, for each t, $\mathbf{H}_t(\theta)$ be a $p \times 1$ vector, a function of observations and parameter θ. Let $\mathbf{H}_t(\theta) = (H_t(\theta, 1), H_t(\theta, 2), \ldots, H_t(\theta, p))'$. The symbol $'$ denotes the transpose of a vector or a matrix. We make the following assumptions. A probability-related or an expectation-related statement holds with respect to the true probability, i.e, P_{θ_0}.

A1. The parameter space Θ is an open set in $\Re^p$, the p-dimensional Euclidean space. The true parameter θ_0 is an interior point of Θ.

A2. $E[H_t(\theta_0, i)] = 0$, for each t and i.

A3. The partial derivatives $\partial H_t(\theta, i)/\partial \theta_j$ exist for each $\theta \in \Theta$ and for each (i, j). Let $D_t(\theta, i, j) = -\partial H_t(\theta, i)/\partial \theta_j$ and let $\mathbf{D}_t(\theta) = ((D_t(\theta, i, j)))$, the $p \times p$ matrix of partial derivatives of $H_t(\theta, i)$ with respect to θ_j's, with a negative sign.

A4. For all i, j, k, $\partial H_t(\theta, i)/\partial \theta_j$ is differentiable with respect to θ_k. Let $\theta_0 = (\theta_{0,1}, \theta_{0,2} \cdots, \theta_{0,p})'$. There exist random variables $M(\theta_0, t)$, such that, in a neighborhood N of θ_0, we have for each t and for each (i, j, k),

$$| \partial^2 H_t(\theta, i)/\partial \theta_k \partial \theta_j | \leq M(\theta_0, t).$$

Further, $E[M(\theta_0, t)]$ is finite for each t.

A5. (i) $\frac{1}{T}\sum_t \mathbf{D}_t(\theta_0) \to \mathbf{D}(\theta_0)$ a.s., where the matrix $\mathbf{D}(\theta_0)$ is a (symmetric) positive definite matrix.

   (ii) $\frac{1}{T}\sum_t \mathbf{H}_t(\theta_0) \to \mathbf{0}$ a.s.

   (iii) There exists a constant M which may depend upon θ_0, such that

$$\frac{1}{T}\sum_t M(\theta_0, t) \to M \quad a.s.,$$

where $0 \le M < \infty$.

(iv) $\frac{1}{\sqrt{T}}\sum_t \mathbf{H}_t(\theta_0) \overset{\mathscr{L}}{\to} N_p(\mathbf{0}, \boldsymbol{\Sigma})$, where the matrix $\boldsymbol{\Sigma}$ is positive-definite.

Theorem 1.5.1 *Under the Assumptions A1–A5 above, we have the following.*

(i) There exists a sequence of estimators $\hat{\theta}_T$ such that

$$P\left[\sum_t \mathbf{H}_t(\hat{\theta}_T) = \mathbf{0}\right] \to 1, \quad and \ \hat{\theta}_T \to \theta_0 \ a.s.$$

(ii) $\sqrt{T}(\hat{\theta}_T - \theta_0) \overset{\mathscr{L}}{\to} N_p(\mathbf{0}, \mathbf{D}^{-1}\boldsymbol{\Sigma}\mathbf{D}^{-1}).$

Proof To prove the above theorem, we begin with a lemma.
Lemma. (Aitchison and Silvey (1958)). If g is a continuous function mapping $\mathfrak{R}^p$ into itself with the property that, for every x such that $\| x \| = 1$, $x'g(0) < 0$, then there exists a point x such that $\| x \| < 1$ and $g(x) = 0$.

Proof of the theorem, part (i). We write $\theta = (\theta_1, \theta_2, \ldots, \theta_p)'$ and $\theta_0 = (\theta_{01}, \theta_{02}, \ldots, \theta_{0p})'$. By the Taylor series expansion of $H_t(\theta, i)$ around the true value θ_0, we have

$$H_t(\theta, i) = H_t(\theta_0, i) - \sum_j D_t(\theta_0, i, j)(\theta_j - \theta_{0,j}) + \frac{1}{2}\sum_j\sum_k (\theta_j - \theta_{0,j})(\theta_k - \theta_{0,k})\frac{\partial^2 H_t(\theta, i)}{\partial\theta_k \partial\theta_j}\bigg|_{\theta^*},$$

where θ^* is an intermediate point on the line segment joining θ and θ_0. We note that $\| \theta^* - \theta_0 \| \le \| \theta - \theta_0 \|$. Now, consider the last term in the above expression. We have, in view of A4,

$$\sum_j\sum_k (\theta_j - \theta_{0,j})(\theta_k - \theta_{0,k})\frac{\partial^2 H_t(\theta, i)}{\partial\theta_k \partial\theta_j}\bigg|_{\theta^*} \le \| \theta - \theta_0 \|^2 \ p^2 M(\theta_0, t).$$

Hence, there exists α such that $| \alpha | \le 1$,

$$\frac{1}{T}H_t(\theta, i) = \frac{1}{T}H_t(\theta_0, i) - \frac{1}{T}\sum_j D_t(\theta_0, i, j)(\theta_j - \theta_{0,j}) + \frac{\alpha p^2}{2} \| \theta - \theta_0 \|^2 \frac{1}{T}M(\theta_0, t).$$

By A5(ii), the first term on the right-hand side converges to 0 a.s., whereas the second term converges to $-\sum_j D(\theta_0, i, j)(\theta_j - \theta_{0,j})$ a.s. Further, since $(1/T)\sum_t M(\theta_0, t)$ a.s. converges to M, the last term eventually does not exceed $(1/2)p^2 \| \theta - \theta_0 \|^2 M$ (with probability one).
Now, since the matrix $D(\theta_0)$ is positive definite, there exists a $\beta > 0$ such that

$(\theta - \theta_0)'D(\theta_0)(\theta - \theta_0) > \beta, \theta \neq \theta_0$. For a given ε, we choose a δ satisfying the following.

  (i) $\delta < \varepsilon$,
  (ii) $\{\theta \mid \parallel \theta - \theta_0 \parallel < \delta\} \subset N$, and
  (iii) $\delta < \beta/[(3/2)p^2(M + 1)]$.

We further choose T large enough such that, for all i, j, k

  (i) $\mid \frac{1}{T}\sum_t H_t(\theta_0, i) \mid \leq \delta^2$,

  (ii) $\frac{1}{T}\sum_t M(\theta_0, t) \leq (M + 1)$,

  (iii) $\mid \frac{1}{T}\sum_t D_t(\theta_0, i, j) - D(\theta_0, i, j) \mid \leq \delta$.
        Then, if $\parallel \theta - \theta_0 \parallel < \delta$, we have

$$\left| \frac{1}{T}\sum_t H_t(\theta, i) + \sum_j D(\theta_0, i, j)(\theta_j - \theta_{0,j}) \right|$$
$$< \delta^2 + \delta p \parallel \theta - \theta_0 \parallel +(1/2)p^2 \parallel \theta - \theta_0 \parallel^2 (M + 1)$$
$$< (3/2)p^2(M + 1)\delta^2.$$

Therefore, if $\parallel \theta - \theta_0 \parallel = \delta$,

$$\frac{1}{T}\sum_i \sum_t (\theta_i - \theta_{0,i})H_t(\theta, i)$$
$$\leq -\sum_i \sum_j D(\theta_0, i, j)(\theta_i - \theta_{0,i})(\theta_j - \theta_{0,j}) + (3/2)p^2(M + 1)\delta^3$$
$$\leq -\beta\delta^2 + (3/2)p^2(M + 1)\delta^3,$$

which is negative in view of the choice of δ, with probability one. By the Aitchison-Silvey Lemma, proof of part (i) is complete.

Proof of part (ii). We have,

$$\sum_t \mathbf{H}_t(\theta) = \sum_t \mathbf{H}_t(\theta_0) - \sum_t \mathbf{D}_t(\theta^*)(\theta - \theta_0)$$

Putting $\theta = \hat{\theta}_T$ in above, we have

$$\sum_t \mathbf{D}_t(\theta_T^*)(\hat{\theta}_T - \theta_0) = \sum_t \mathbf{H}_t(\theta_0).$$

Since $\sum_t \mathbf{D}_t(\theta_0)/T$ in non-singular for a large T (with probability one) and $\sum_t \mathbf{D}_t(\theta)/T$ is continuous, it follows that $\sum_t \mathbf{D}_t(\theta_T^*)/T$ is non-singular with probability one. Moreover, if θ_T^* is an intermediate point,

$$\sqrt{T}(\hat{\theta}_T - \theta_0) = \left[\frac{1}{T}\sum_t \mathbf{D}_t(\theta_T^*)\right]^{-1} \frac{1}{\sqrt{T}}\sum_t \mathbf{H}_t(\theta_0).$$

It suffices to show that $(1/T)\sum_t \mathbf{D}_t(\theta_T^*) \to \mathbf{D}(\theta_0)$ a.s. We note that

$$\left\|\frac{1}{T}\sum_t \mathbf{D}_t(\theta_T^*) - \frac{1}{T}\sum_t \mathbf{D}_t(\theta_0)\right\| \leq \frac{1}{T}\sum_t M_t(\theta_0)\,\|\,\hat{\theta}_T - \theta_0\,\|\,.$$

By A5(iii), $(1/T)\sum_t M_t(\theta_0) \to M$ a.s. which is finite and
$\|\,\hat{\theta}_T - \theta_0\,\| \to 0$ a.s. in view of consistency of $\hat{\theta}_T$. Thus,

$$\sqrt{T}(\hat{\theta}_T - \theta_0) - [\mathbf{D}(\theta_0)]^{-1}\sum_t \mathbf{H}_t(\theta_0)/\sqrt{T} \to \mathbf{0} \quad \text{a.s.}$$

The asymptotic normality follows in view of A5 (iv).

The above proof follows Billingsley (1961a), who deals with theory of likelihood equations obtained from stationary ergodic Markov sequences , see Rajarshi (1987), who deals with general random sequences.
      The condition A5(i) may look rather restrictive, but actually, it holds for many methods of estimation such as maximum likelihood, conditional least squares (which is discussed in Chap. 4). If the functions H_t's are themselves obtained as equations for minimizing or maximizing a function, the condition of symmetry of D in A5 (i) is the same as interchangeability of partial derivatives with respect to θ_i and θ_j.
      Let $S_t(\theta, i) = \frac{\partial \ln f(X_t|X_{t-1}, X_{t-2}, \ldots, X_0)}{\partial \theta_i}$, the ith component of the vector-valued score function. In the maximum likelihood method, we take $H_t(\theta, i) = S_t(\theta, i)$. We note that $E[H_t(\theta, i) \mid X_t, X_{t-1}, \ldots, X_0] = 0$, if the Cramér regularity conditions hold for the conditional density. Thus, $\sum H_t(\theta, i)$ is a martingale for each i. The Central Limit Theorem as required in A5 (iv) follows by the Billingsley-Ibragimov Martingale CLT . The almost sure convergence as required in Assumptions, follows from the Ergodic theorem. We also note that the (i, j)th element of $I(\theta)$, the Fisher Information matrix is given by

$$[I(\theta)]_{ij} = \lim_{T \to \infty} \frac{1}{T}\sum_t E[S_t(\theta, i)S_t(\theta, j)].$$

Moreover, in view of the regularity conditions, $\boldsymbol{\Sigma} = I(\theta) = \mathbf{D}$, the matrix of expectations of second order derivatives. We refer to Basawa and Prakasa Rao (1980), Chap. 7 for discussion of properties of maximum likelihood estimators obtained from a general random sequence. In the case of stationary Markov sequences of order 1, this reduces to

$$I(\theta) = \left(\left(E \left[\frac{\partial \ln f_\theta(X_t \mid X_{t-1})}{\partial \theta_i} \frac{\partial \ln f_\theta(X_t \mid X_{t-1})}{\partial \theta_j} \right] \right) \right).$$

If $E[H_t(\theta, i) \mid X_t, X_{t-1}, \ldots, X_0] \neq 0$, one cannot apply Martingale CLT. The information about the parameters is based on a finite dimensional marginal distribution. If the underlying sequence of observations is strong mixing and if H_t is a function of X_t (and θ), the CLT (Theorem 1.3.4) for sums of functions of strong mixing random variables, can be applied. In the simplest case, the sample mean has been shown to be a $\sqrt{T}$-CAN estimator . Results for Fréchet differentiable functions, which we have stated earlier, do not follow from the above theorem. In general, the CAN property of estimators obtained as solutions of some equations also does not follow from the above theorem. We state a theorem below, which covers a class of robust estimators frequently known as Generalized M estimators .

Let the state-space of the observable sequence $\mathfrak{X}$ be $\mathfrak{R}^d$. Let a $p \times 1$ parameter be defined as the unique solution of the equation $E[\psi(X_1, X_2, \ldots, X_m, \theta)] = 0$, where ψ is a function from $\mathfrak{R}^{md} \times \mathfrak{R}^p$ to $\mathfrak{R}^p$. An M estimator is defined as a solution of the following equation in θ.

$$\frac{1}{T - m + 1} \sum_{t=1}^{T-m+1} \psi(X_t, X_{t+1}, \ldots, X_{t+m-1}, \theta) = 0.$$

We recall that a function $g : \mathfrak{R} \to \mathfrak{R}$ is said to satisfy a Lipschitz condition of order κ, if $|g(x) - g(y)| \leq C|x - y|^\kappa$, for some constant C. Let D^α denote the differential operator $D^\alpha = \frac{\partial^{\alpha_1 + \alpha_2 + \cdots + \alpha_k}}{\partial x_1^{\alpha_1} \cdots \partial x_k^{\alpha_k}}$ on $\mathfrak{R}^k$. For a vector x, $\| x \| = (\sum x_i^2)^{1/2}$. For a matrix A, $\| A \|$ denotes $sup\{\| Ax \| \mid \| x \| = 1\}$.

Theorem 1.5.2 *(Theorem 4.2 of Lahiri 2003) We assume that the above equation has a unique solution $\hat{\theta}$, which is a random variable. Let $y = (x_1, x_2, \ldots, x_m)$ and for all t, $Y_t = (X_t, X_{t+1}, \ldots, X_{t+m-1})$. We assume the followings.*

A1. *The function $\psi(y, \theta)$ is differentiable with respect to θ almost everywhere under μ_F, the measure induced by the distribution function $F(y)$ of Y.*

A2. *The partial derivatives of ψ with respect to θ_i's satisfy a Lipschitz condition of order κ a.e. μ_F, where $0 < \kappa < 1$.*

A3. $E(Y, \theta) = 0$.

A4. *Let $\Sigma = \lim_{T_0 \to \infty} Var\left(T_0^{-1/2} \sum_{t=1}^{T_0} \psi(Y_t, \theta) \right)$, where $T_0 = T + m - 1$. Then, Σ is a positive definite matrix.*

A5. *Let the matrix $\mathbf{D}$ be defined by $E\left[\frac{\partial \psi_i(Y_1, \theta)}{\partial \theta_j} \right]$ as its (i, j)th element. Then, the matrix $\mathbf{D}$ is non-singular.*

A6. *There exists a $\delta > 0$ such that $E\left[\| D^\alpha \psi(Y_1, \theta) \|^{2+\delta} \right] < \infty$ for all α such that $\sum \alpha_i = 0, 1$ and $\alpha_i \geq 0$ for each i. Further, the sequence $\mathfrak{X}$ is strong mixing with $\sum_{t=1}^\infty t[\alpha(t)]^{\delta/(2+\delta)} < \infty$.*

Then, $\hat{\theta} \to \theta$ in probability and $\sqrt{T}(\hat{\theta} - \theta) \xrightarrow{\mathscr{D}} N_p(0, \mathbf{D}^{-1}\Sigma\mathbf{D}^{-1'})$.

Theorems in this chapter lead to CAN estimators for parameters of various stochastic models discussed in the Chaps. 2, 3 and 4. Theorems 1.2.2, 1.5.1, and 1.5.2 can be employed to generate CAN estimators in semi-parametric and parametric models, where likelihood or some partial information is available. Theorems 1.3.4, 1.4.1, and 1.4.2 lead to CAN property for estimators in situations, where apart from mixing and stationarity properties, very few assumptions have been made regarding the joint or marginal distributions of observations.

References

Aitchison, J., Silvey, S.D.: Maximum-likelihood estimation of parameters subject to restraints. Ann. Math. Statist. **29**, 813–828 (1958)

Athreya, K.B., Lahiri. S.N.: Measure Theory and Probability. Springer, New York (2006)

Athreya, K.B., Pantula, G.S.: Mixing properties of Harris chains and Auto-regressive Processes. J. Appl. Probab. **23**, 880–892 (1986a)

Athreya, K.B., Pantula, G.S.: A note on strong mixing ARMA processes. Statist. Probab. Lett. **4**, 187–190 (1986b)

Basawa, I.V., Prakasa Rao, B.L.S.: Statistical Inference for Stochastic Processes. Academic Press, London (1980)

Bickel, P.J., Bühlmann, P.: A new mixing notion and functional central limit theorems for a sieve bootstrap in time series. Bernoulli. **5**, 413–446 (1999)

Billingsley, P.: Statistical Inference for Markov Processes. The University of Chicago Press, Chicago (1961a)

Billingsley, P.: The Lindeberg-Lévy theorem for martingales. Proc. Am. Math. Soc. **12**, 788–792 (1961b)

Billingsley, P.: Convergence of Probability Measures. Wiley, New York (1968)

Boos, D.D.: A Differential for L Statistics. Ann. Statist. **7**, 955–959 (1979)

Bosq, D.: Nonparametric Statistics for Stochastic Processes Lecture Notes in Statistics 110. Springer, New York (1996)

Bradley, R.A.: Basic properties of strong mixing conditions : A survey and some open questions. Probab. Surv. **2**, 107–144 (2005)

Bühlmann, P.: Blockwise bootstrapped empirical process for stationary sequences. Ann. Statist. **22**, 995–1012 (1994)

Deo, C.M.: A note on empirical processes of strong mixing sequences. Ann. Probab. **1**, 870–875 (1973)

Doukhan, P.: Mixing : Properties and examples lecture notes in statistics 85. Springer, New York (1994)

Fernholz, L.T.: Von Mises Calculus for Statistical Functionals. Lecture Notes in Statistics, Vol. 19, Springer, New York (1983)

Gorodetskii, V.V.: On strong mixing property for linear sequences theory. Probab. Appl. **22**, 411–413 (1977)

Götze, F., Hipp, C.: Asymptotic expansions for sums of weakly dependent random variables. Z. Wahr. verw. Geb. **64**, 211–240 (1983)

Hall, P., Heyde, C.C.:Martingale Limit Theory and its Applications Academic Press, London (1980)

Ibragimov, I.A.: A central limit theorem for a class of dependent random variables theory. Probab. Appl. **8**, 83–89 (1963)

Lahiri, S.N.: Resampling Methods for Dependent Data. Springer, New York (2003)

Lindsey, J.K.: Statistical Analysis of Stochastic Processes in Time. Cambridge University Press, New York (2004)

Nze, P.A., Bühlmann, P., Doukhan, P.: Weak Dependence beyond mixing and asymptotics for nonparametric regression. Ann. Statist. **30**, 397–430 (2002)

Nze, P.A., Doukhan, P.: Weak dependence : models and applications. In: Dehling, H., Mikosch, T., Sørensen, M. (eds.) Empirical Process Techniques for Dependent Data, pp. 117–136. Birkhäuser, Boston (2002)

Radulović, D.: On the Bootstrap and the Empirical Processes for Dependent Sequences. In: Dehling, H., Mikosch, T., Sørensen, M. (eds.) Empirical Process Techniques for Dependent Data, pp. 345–364, Birkhäuser, Boston (2002)

Rajarshi, M.B.: Chi-squared type goodness of fit tests for stochastic models through conditional least squared estimation. J. Ind. Statist. Assoc. 24, 65–76 (1987)

Serfling, R.J.: Approximation Theorems of Mathematical Statistics. Wiley, New York (1980)

Shao, J.: Mathematical Statistics. Springer, New York (1999)

Tjøstheim, D.: Non-linear Time Series: A Selective Review Scand. J. Statist. **21**, 97–130 (1994)

Withers, C.S.: Convergence of empirical processes of mixing rv's on [0,1]. Ann. Statist. **3**, 1101–1108 (1975)

Withers, C.S.: Conditions for linear processes to be strong mixing. Z. Wahr. Verw. Geb. **57**, 477–480 (1981)

Yoshihara, K.: Weak convergence of multidimensional empirical processes for strong mixing sequences of stochastic vectors. Z. Wahr. Verw. Geb. **33**, 133–137 (1975)

Chapter 2
Markov Chains and Their Extensions

Abstract This chapter deals with likelihood-based inference for ergodic finite as well as infinite Markov chains. We also consider extensions of Markov chain models, such as Hidden Markov chain, Markov chains based on polytomous regression, and Raftery's Mixture Transition Density model. These models have less number of parameters as compared to a higher order finite Markov chain. Lastly, we discuss methods of estimation in grouped data from finite Markov chains.

2.1 Markov Chains

Let $\mathfrak{X} = (X_0, X_1, \ldots)$ be an M-state Markov chain with the state-space $S = \{1, \ldots, M\}$ and the one-step transition probability matrix (t.p.m.) $P = ((p_{ij}))_{M \times M}$, $p_{ij} \geq 0, \forall (i, j)$ and $\sum_{j=1}^{M} p_{ij} = 1, \forall i$. We assume that all states communicate with each other and are aperiodic, which implies that all states are non-null persistent and hence the chain is ergodic. Let $\{\pi_i, i = 1, 2, \ldots, M\}$ be the unique stationary distribution of the Markov chain $\mathfrak{X}$. It then follows that for all i, j,

$$p_{ij}^{(t)} \to \pi_j > 0, \text{ as } t \to \infty.$$

Let $(X_0, X_1, \ldots, X_T)$ be the $T + 1$ successive observations from the above Markov chain. The conditional likelihood (given the initial observation X_0) is given by $L(P) = \prod_{t=0}^{T} p_{x_t, x_{t+1}}$. Thus,

$$\ln(L(P)) = \sum_i \sum_j N_{ij} \ln p_{ij} \quad N_{ij} = \sum_{t=0}^{T-1} I[X_t = i, X_{t+1} = j].$$

We need to maximize $\ln(L(P))$ with respect to p_{ij}'s, subject to the constraints that $\sum_j p_{ij} = 1, \forall i$. Let λ_i's be the Lagrangian parameters. We set

M. B. Rajarshi, *Statistical Inference for Discrete Time Stochastic Processes*,
SpringerBriefs in Statistics, DOI: 10.1007/978-81-322-0763-4_2,
© The Author(s) 2012

$$f = \ln(L(P)) + \sum_{i=1}^{M} \lambda_i \left(\sum_j p_{ij} - 1 \right).$$

Then, if $p_{ij} > 0$, it can be easily shown that

$$\hat{p}_{ij} = \frac{N_{ij}}{N_{i+}}, \quad \text{where } N_{i+} = \sum_j N_{ij}.$$

Let $Z_t = I[X_t = i, X_{t+1} = j]$. We note that Z_t is a Bernoulli random variable. Let $\{p_i, i = 1, 2, \ldots, M\}$ be the initial distribution of $\mathfrak{X}$. Then,

$$E[N_{ij}] = \sum_{t=0}^{T-1} E[Z_t] = \sum_{t=1}^{T-1} \sum_{k=1}^{M} \left(p_k p_{ki}^{(t)} p_{ij} \right).$$

Since $p_{ki}^{(t)} \to \pi_i$, $\sum_{t=1}^{T-1} p_{ki}^{(t)}/T \to \pi_i$. Thus, $E[N_{ij}/T] \to \pi_i p_{ij}$. Further, we observe that

$$\text{Var}[N_{ij}] = \sum_{t=1}^{T-1} \sum_{k=1}^{M} \left(p_k p_{ki}^{(t)} p_{ij} \right) \left(1 - \left(p_k p_{ki}^{(t)} p_{ij} \right) + \sum_{k=1}^{M} \sum_{s \neq t} \left(p_k p_{ki}^{(s)} p_{ij} \right) p_{ki}^{(t)} p_{ij} \right)$$

and that $V[N_{ij}/T] \to 0$ as $T \to \infty$. It follows that, for all i, j, $N_{ij}/T \to \pi_i p_{ij}$ in probability and hence $\hat{p}_{ij} \to p_{ij}$ in probability. We observe that the above convergence holds for any initial distribution $\{p_i, i = 1, 2, \ldots, M\}$. It is easy to derive the Fisher information matrix from the above likelihood. It follows from Theorem 1.5.1 (and also from Theorem 1.1 of Billingsley 1961) that the joint distribution of $\sqrt{T}(\hat{p}_{ij} - p_{ij})$ $i, j \in S$ (written in a suitable vector form) is asymptotically multivariate normal with the mean vector 0 and variances and covariances given by

$$V(\sqrt{T}(\hat{p}_{ij} - p_{ij})) \approx \pi_i p_{ij}(1 - p_{ij})$$

$$\text{Cov}(\sqrt{T}(\hat{p}_{ij} - p_{ij}), \sqrt{T}(\hat{p}_{ik} - p_{ik})) \approx -\pi_i p_{ij} p_{ik}, \, j \neq k.$$

$$\text{Cov}(\sqrt{T}(\hat{p}_{ij} - p_{ij}), \sqrt{T}(\hat{p}_{lk} - p_{lk})) \approx 0, \quad if \, i \neq l.$$

It may be remarked that the likelihood function and hence the variance-covariance matrix resemble the likelihood function and variance-covariance pattern of M independent multinomial distributions respectively.

Goodness of fit of a Markov chain
1. LRT and Pearson's X^2 statistics for testing the order. The first order Markov property can be judged by testing against the second order. Under the assumption of the second order, the ML estimate of $p_{ij,k} = P[X_{t+2} = k | X_t = i, X_{t+1} = j]$ is

given by $\hat{p}_{ij,k} = \frac{N_{ijk}}{N_{ij+}}$, where $N_{ijk} = \sum_{t=1}^{T-2} I[X_t = i, X_{t+1} = j, X_{t+2} = k]$ and $N_{ij+} = \sum_k N_{ijk}$. We assume that the vector valued Markov chain $\{(X_t, X_{t+1}), t \geq 0\}$ is irreducible and aperiodic. Thus, the LRT statistic is given by

$$2\left\{\sum_{ijk} N_{ijk} \ln \hat{p}_{ijk} - \sum_{ij} N_{ij} \ln \hat{p}_{ij}\right\} \sim \chi^2_{(M-1)^2 M},$$

as the difference in the numbers of parameters of the two models is $M^2(M - 1) - M(M - 1) = (M - 1)^2 M$. The corresponding Pearson's X^2-statistic is given by

$$X^2 = \sum_{ijk} \frac{(N_{ijk} - E_{ijk})^2}{E_{ijk}} \quad \text{where} \quad E_{ijk} = N_{i++}\hat{p}_{ij}\hat{p}_{jk}.$$

The degrees of freedom are the same as that of the LRT. One can similarly obtain ML estimates of an L order Markov chain and develop a large sample chi-squared test for an L-th order Markov chains against a Markov chain of order $(L + 1)$.

2. *Estimation of the true order of a Markov chain (Selection of a Markov model).* The above method of testing an order L against $L + 1$ is sequential in nature. We continue to test until an r-order model is not rejected against the $(r + 1)$-th order model. Since this procedure involves a series of tests to be carried out in a sequential manner, the probability of selecting the underlying true model (of unknown order) may not asymptotically converge to 1. A better and theoretically valid procedure is to apply the information criteria such as akaike's information criterion (AIC) or bayes information criterion (BIC) . These are defined below. Let K be the number of parameters of a model under consideration.

$$\text{AIC} = 2 \sup_\theta \ln L(\theta) - T,$$

$$\text{BIC} = 2 \sup_\theta \ln L(\theta) - K \ln(T).$$

Both the log-likelihood and the number K of parameters change from model to model. We select the model which has the smallest AIC/BIC. Katz (1981) has shown that the BIC procedure gives a consistent estimator of the true order of a Markov model. The AIC frequently overestimates the true order. Another advantage of both AIC and BIC is that they can be applied even if the models are not nested. A construction of LRT requires the nested property, e.g., the L-th order Markov model is included in the $(L + 1)$-th order Markov model.

3. *Time homogeneity.* The assumption of time homogeneity can be tested by dividing the data into several non-overlapping blocks of consecutive observations of moderate length. We then compute $\hat{p}_{ij}$ for each (i, j) obtained from each of such blocks and plot the estimators against the time. If the transition probabilities vary a lot over

time, such plots would reveal changes. (A formal test is difficult to construct.) The procedure can be carried out based on overlapping blocks also.

4. *Distribution of patterns.* Application of a two-state Markov chain to a sequence of dry (the state 1) and wet (rainy) (the state 0) days over a monsoon at a place requires fitting of distributions of lengths of wet and dry cycles or spells. Length of a wet cycle equals r with probability $p_{00}{}^{r-1}p_{01}$. Observed distribution of wet cycle lengths can be compared with the distribution with estimated parameters. Visual inspection of plots suffices in most of the cases. The Pearson's X^2 statistic can be computed while comparing with other models. It may be pointed out that the X^2 statistic does not have a chi-square distribution. We refer to Table 2.10 (p. 76) of Guttorp (1995) for such an application.

2.2 Parametric Models

A parametric model with M states for $\mathfrak{X}$ is given by specifying p_{ij} as functions of θ, a p-dimensional parameter, i.e., as $p_{ij}(\theta) = P_\theta[X_{t+1} = j | X_t = i]$. The parameter space is Θ, an open subset of p-dimensional Euclidean space. We make the following assumptions.

1. The set $D = \{(i, j) | p_{ij}(\theta) > 0\}$ does not depend upon θ.
2. Each of the functions $p_{ij}(\theta)$ is twice differentiable with respect to θ_r, $r = 1, 2, \ldots, p$. The second derivatives are continuous.
3. The appropriately constructed matrix $\frac{\partial p_{ij}(\theta)}{\partial \theta_r}$, $(i, j) \in D$, $r = 1, 2, \ldots, p$ of order $d \times p$ is of rank p where d is the number of elements in the set D.
4. $\forall \theta$, $\mathfrak{X}$ is an irreducible, non-null persistent, aperiodic Markov chain .

The above assumptions can be described as extensions of Cramer regularity conditions, assumed usually in the case of i.i.d. observations. Let $\pi_i(\theta)$ denote the unique stationary distribution of $\mathfrak{X}$. For the time being, we assume that the chain is in equilibrium, i.e., X_0 follows the distribution $\pi_i(\theta)$. It is easily seen that $\ln L(\theta) = \sum_{i,j} N_{ij} \ln p_{ij}(\theta) + \sum_i I[X_0 = i]\pi_i(\theta)$. We ignore the second term as before. Consequently, the likelihood equations are given by

$$\frac{\partial \ln L(\theta)}{\partial \theta_r} = \sum_{(i,j) \in D} \frac{\partial p_{ij}(\theta)}{\partial \theta_r} \frac{N_{ij}}{p_{ij}(\theta)} = 0, \quad r = 1, 2, \ldots, p.$$

Further,

$$\frac{\partial^2 \ln L(\theta)}{\partial \theta_r \partial \theta_s} = \sum_{(i,j) \in D} \left[\frac{N_{ij}}{p_{ij}(\theta)} \frac{\partial^2 p_{ij}(\theta)}{\partial \theta_r \partial \theta_s} - \frac{N_{ij}}{(p_{ij}(\theta))^2} \frac{\partial p_{ij}(\theta)}{\partial \theta_r} \frac{\partial p_{ij}(\theta)}{\partial \theta_s} \right].$$

Now, since $N_{ij} = \sum_{t=0}^{T-1} I[X_t = i, X_{t+1} = j]$, in view of stationarity, we have

$$E(N_{ij}) = T\pi_i(\theta)p_{ij}(\theta).$$

Since $\sum_j p_{ij}(\theta) = 1$, $\sum_{(i,j)\in D} \pi_i \frac{\partial^2 p_{ij}(\theta)}{\partial\theta_r \partial\theta_s} = 0$ in view of the regularity conditions. The Fisher Information Matrix is defined by

$$I(\theta) = \left(\left(\lim_{T\to\infty} \frac{1}{T} E\left[\frac{\partial \ln L(\theta)}{\partial\theta_r} \frac{\partial \ln L(\theta)}{\partial\theta_s} \right] \right)\right).$$

In view of the regularity conditions, it is then given by

$$I(\theta) = ((I_{rs}(\theta))),$$

where

$$I_{rs}(\theta) = - \sum_{(i,j)\in D} \frac{\pi_i(\theta)}{p_{ij}(\theta)} \frac{\partial p_{ij}(\theta)}{\partial\theta_r} \frac{\partial p_{ij}(\theta)}{\partial\theta_s}.$$

The expected values in the above expressions have been derived with respect to the joint distribution of (X_0, X_1) under the assumption of stationarity. Let $\hat{\theta}$ be a consistent solution of the likelihood equations. It follows that

$$\sqrt{T}(\hat{\theta} - \theta) \xrightarrow{\mathcal{D}} N_p \left(\mathbf{0}, (I(\theta))^{-1} \right).$$

An estimator of the Fisher Information matrix is needed in construction of confidence regions and tests of hypotheses. In practice, it is easier to use the observed Fisher Information matrix, whose (i, j)-th element is given by

$$F_{ij}(\hat{\theta}) = - \left. \frac{\partial^2 \ln L(\theta)}{\partial\theta_r \partial\theta_s} \right|_{\theta=\hat{\theta}}$$

and the corresponding estimator of $I(\theta)$ is given by

$$\hat{I}(\theta) = \left(\left(F_{ij}(\hat{\theta})/T \right)\right).$$

Another estimator of $I(\theta)$ is $I(\hat{\theta})$, which is obtained by replacing θ by $\hat{\theta}$ in $I(\theta)$. However, this requires a theoretical derivation of the expectations involved, which could be tedious in some cases.

Goodness of fit of parametric finite Markov chain. models: This is similar to goodness-of-fit procedures for parametric multinomial models. We note that under H_0, $\mathfrak{X}$ follows the above parametric model, $E_{H_0}(N_{ij}) = T\pi_i(\theta)p_{ij}(\theta)$, the estimate of which is given by $E_{ij} = N_{i+}p_{ij}(\hat{\theta})$, under H_0. Thus, the Pearsonian X^2 statistic is given by

$$X^2 = \sum_{(i,j) \in D} \frac{\left(N_{ij} - N_{i+} p_{ij}(\hat{\theta})\right)^2}{N_{i+} p_{ij}(\hat{\theta})}$$

and its asymptotic null distribution is χ^2 with (the number of elements in the set D) $-$ $M - p$ as the degrees of freedom. The LRT statistic has the same asymptotic distribution, under H_0.

One can consider the Q–Q plot of $\left(N_{ij} - N_{i+} p_{ij}(\hat{\theta})\right) \big/ \sqrt{N_{i+} p_{ij}(\hat{\theta})}$ for $(i, j) \in D$ by regrading them observations from $N(0, 1)$. Such a plot may reveal cells which have a sizable contribute to the LRT or X^2, cf. Davison (2003), p. 235.

Testing for sub-models. Under the above stated conditions, we also get the chi-square distribution of the LRT for the null hypothesis $H_0 : \theta \in \Theta_0$, provided that the regularity conditions hold for the parameter space Θ_0. The degrees of freedom for the chi-square are given by the $p - p_0$, where p_0 is the number of (distinct) parameters corresponding to H_0.

Example 2.2.1 Consider the two-state Markov chain with its t.p.m. given by

$$\begin{array}{cc} & \begin{array}{cc} 0 & 1 \end{array} \\ \begin{array}{c} 0 \\ 1 \end{array} & \begin{pmatrix} 1 - \theta & \theta \\ \theta & 1 - \theta \end{pmatrix} \end{array}.$$

The null parameter space is $\Theta_0 = (0, 1)$ and for all θ, $D = \{(0, 0), (0, 1), (1, 0), (1, 1)\}$. The vector of derivatives of the elements of the t.p.m., taken row-wise is given by $(-1, 1, 1, -1)$, whose rank is 1. Besides, the chain is irreducible and aperiodic with $(1/2, 1/2)$ as the unique stationary distribution. It is straightforward that the MLE is given by $\hat{\theta} = (N_{01} + N_{10})/T$. The asymptotic null distribution of each of the statistics X^2 and LRT is χ_1^2.

Example 2.2.2 Let $\Theta = (0, 1)$ and let the t.p.m. of a Markov chain be given by

$$\begin{pmatrix} 1 & 0 \\ \theta & 1 - \theta \end{pmatrix}.$$

The Markov chain is reducible and not ergodic. One can directly study the behavior of the MLE and verify that it is not consistent, for any initial distribution.

Example 2.2.3 The hypothesis that $\mathfrak{X}$ is a sequence of i.i.d. random variables within the hypotheses that $\mathfrak{X}$ is a first order Markov chain can be represented as a parametric model where $p_{ij} = \pi_j$ for all i, j, where $\sum_j \pi_j = 1$. Obviously, $\hat{\pi}_j = N_j/T$. The LRT and the Pearson's X^2 statistics are respectively given by

$$-2 \ln \Lambda = -2 \sum_{i,j} \ln \left(\frac{T N_{ij}}{N_i N_j}\right)$$

$$X^2 = \sum_{i,j} \frac{(N_{ij} - N_i N_j / T)^2}{N_i N_j / T}.$$

Under H_0, both have a chi-square distribution with $M(M-1) - M - 1 = (M-1)^2$ d.f. Similarity with testing for independence in a contingency table is obvious.

Example 2.2.4 Let us suppose that $\mathfrak{X}$ is a second order Markov chain. (For verification of various conditions, it may be noted that a second order Markov chain can be represented as a first order vector-valued Markov chain $\{Y_t, t \geq 0\}$ where $Y_t = (X_t, X_{t+1})$.) Let us assume that $p_{ijk} > 0$ for all i, j, k. The hypothesis that $\{X_t, t \geq 0\}$ is a first order Markov chain represents a parametric model with $p_{ijk} = p_{jk}$ for all i, j, k. This justifies the large sample distributions of the two statistics for testing the first order against (*within*) the second order, stated earlier.

Example 2.2.5 Consider the Markov chain with the t.p.m.

$$
\begin{array}{c@{\quad}ccccc}
 & 1 & 2 & 3 & 4 & 5 \\
1 & \left(\begin{array}{c}0\end{array}\right. & 3/4 & 0 & 1/4 & 0 \\
2 & 0 & \theta & 1-\theta & 0 & 0 \\
3 & 0 & \theta/2 & 1-\theta/2 & 0 & 0 \\
4 & 0 & 0 & 0 & 1-\theta^2 & \theta^2 \\
5 & 0 & 0 & 0 & \theta & \left.1-\theta\end{array}\right),
\end{array}
$$

where $0 < \theta < 1$. It is easy to see that there are two persistent and minimal closed classes viz., $\{2, 3\}$ and $\{4, 5\}$. The state 1 is transient. The regularity conditions in terms of continuity and differentiability are satisfied. But the probabilistic conditions of a parametric model are not satisfied. It can be shown that the MLE is consistent, however its asymptotic distribution is a mixture of two normal distributions, if $X_0 = 1$.

Example 2.2.6 Testing for a specified stationary distribution

The hypothesis of interest is $H_0 : \pi_i = \pi_i(0)$ where $\pi_i(0)$, $i = 1, 2, \ldots, M$ is the specified stationary distribution of the Markov chain. We construct the LRT for this problem.
Consider the simple case $M = 2$ first. Let $(\pi_0(0), \pi_1(0))$, with $\pi_0(0) + \pi_1(0) = 1$, be the stationary distribution under H_0. For the sake of notational convenience, let us denote the specified stationary distribution by (π_0, π_1). The log-likelihood

$$N_{00} \ln p_{00} + N_{01} \ln p_{01} + N_{10} \ln p_{10} + N_{11} \ln p_{11}$$

needs to be maximized subject to the constraints that
(1) $\pi_0 p_{00} + (1 - \pi_0) p_{10} = \pi_0$
(2) $\pi_0(1 - p_{01}) + (1 - \pi_0)(1 - p_{11}) = 1 - \pi_0$.

However, since $p_{00} + p_{01} = 1$ and $p_{10} + p_{11} = 1$, we need to consider only one of these two constraints. We take the constraint (1). By the Lagrange's multiplier theorem, we set

$$g = N_{00} \ln p_{00} + N_{01} \ln(1 - p_{00}) + N_{10} \ln p_{10} + N_{11} \ln(1 - p_{10})$$
$$+ \lambda(\pi_0 p_{00} + (1 - \pi_0)p_{10} - \pi_0).$$

Then, we have the following equations to get ML estimator under H_0

$$\frac{\partial g}{\partial p_{00}} = \frac{N_{00}}{p_{00}} - \frac{N_{01}}{1 - p_{00}} + \lambda\pi_0 = 0.$$
$$\frac{\partial g}{\partial p_{10}} = \frac{N_{10}}{p_{10}} - \frac{N_{11}}{1 - p_{10}} + (1 - \pi_0)\lambda = 0.$$
$$\frac{\partial g}{\partial \lambda} = \pi_0 p_{00} + (1 - \pi_0)p_{10} - \pi_0 = 0.$$

The above system can be solved by an iterative scheme. For an M state Markov chain, we set

$$g = \sum_{i,j=1}^{M} N_{ij} \ln p_{ij} + \sum_{i=1}^{M-1} \lambda_i \left(\pi_{0i} - \sum_{j=1}^{M} \pi_{0j} p_{ji}\right) + \sum_{i=1}^{M} \eta_i \left(\sum_{j=1}^{M} p_{ij} - 1\right),$$

where λ_i's and η_j's are the Lagrangian parameters. The LRT has a large sample χ^2_{M-1} distribution, under H_0.

Another strategy is to construct the Wald's test . We compute the stationary distribution of $\hat{P}$, the MLE of P and compare it with $\pi_i(0)$. Let $\hat{\pi}_i$, $i = 1, 2, \ldots, M$ be the stationary distribution of $\hat{P}$. To construct the Wald's test for H_0, we need to compute the variance-covariance matrix of $\hat{\pi}_i$, $i = 1, 2, \ldots, M$.

There are three vectors to be compared: the observed proportions of various states, $\hat{\pi}_i$, $i = 1, 2, \ldots M$ (the stationary distribution of $\hat{P}$) and $\pi_i(0)$. The first vector is almost the same as the stationary distribution of $\hat{P}$ for a large T.

It is more convenient to construct a Wald's test based on a quadratic form in observed proportions of various states. Let $U_t(i) = I[X_t = i]$, $i = 1, 2, \ldots, M$, $t = 1, 2, \ldots, T$. Then, $\tilde{\pi}_i = \bar{U}_T(i) = \sum_{t=1}^{T} I[X_t = i]/T$. Let $p_{ij}^{(t)} = P[X_t = j \mid X_0 = i]$ denote a t-step transition probability. Then,

$$\text{Cov}\left(\bar{U}_T(i), \bar{U}_T(j)\right) = \frac{1}{T^2} \sum_{s=1}^{T} \sum_{t=1}^{T} \text{Cov}(U_s(i), U_t(j))$$
$$= \frac{1}{T^2} \sum_{s=1}^{T} \sum_{t=1}^{T} \left[\pi_i p_{ij}^{(|t-s|)} - \pi_i \pi_j\right].$$

The stationary probabilities can be estimated by the observed proportions $\tilde{\pi}_i$'s. Powers of $\hat{P}$ can be used to estimate $p_{ij}^{(t)}$'s in the above.

Another and possibly simpler procedure to estimate the above variances and covariances is as follows. In view of the stationarity, the covariance $\text{Cov}(U_s(i), U_t(j))$ is a function of $|t - s|$ only. Consider then $\text{Cov}(U_1(i), U_d(j)), d \geq 1$ which can be estimated by

$$\hat{\text{Cov}}(U_1(i), U_d(j)) = \frac{1}{T - d - L} \sum_{s=1}^{T-d-L} \left(U_s(i) - \bar{U}(1, i)\right) \left(U_{s+d+L}(j) - \bar{U}(2, j)\right),$$

(2.1)

where $\bar{U}(1, i) = \frac{1}{T-d-L} \sum_{s=1}^{T-d-L} U_s(i)$ and $\bar{U}(2, j) = \frac{1}{T-d-L} \sum_{s=1}^{T-d-L} U_{s+d+L}(j)$ and $L = L(T)$ is a sequence of integers such that $L \to \infty$ and $\sqrt{L}/T \to 0$. In practice, for large T, the two means $\bar{U}(1, i)$ and $\bar{U}(2, j)$ can be replaced by the mean of all the observations, however, in such a case, it is possible that for some samples, the corresponding estimator of the variance-covariance matrix of $\left(\bar{U}_T(1), \bar{U}_T(2), \ldots, \bar{U}_T(M)\right)$ is not non-negative definite. We notice that the above estimator does not depend on the assumed Markov model and it can be shown to be consistent for sequences more general than Markov chains. Under the assumption that the covariances of larger lags are negligible, it focuses on the more dominant terms of the covariance. It is a special case of estimators that we discuss in some more detail in Sect. 6.6. Under the assumptions that we have made, the Markov chain is geometrically ergodic (i.e., $|p_{ij}^{(t)} - \pi_j| < C\rho^t, 0 \leq \rho < 1$). As remarked in Example 1.3.2, it is Geometrically Strong Mixing (It can be shown that such an estimator of the variance is consistent, cf. Theorem 6.6.1).

Let $\hat{\Sigma}$ be a consistent estimator of the variance covariance of the vector of observed proportions of states $\tilde{\Pi} = (\tilde{\pi}_1, \tilde{\pi}_2, \ldots, \tilde{\pi}_M)'$ and let $\Pi(0) = (\pi_1(0), \pi_2(0), \ldots, \pi_M(0))'$. Then, the Wald test-statistic is given by

$$X^2 = T \left(\tilde{\Pi} - \Pi(0)\right)' \hat{\Sigma}^+ \left(\tilde{\Pi} - \Pi(0)\right),$$

(2.2)

where A^+ denotes the Moore-Penrose g-inverse of a matrix A. Under H_0, the test-statistic X^2 has asymptotically a χ_{M-1}^2 distribution under H_0.

Markov chains with infinitely many states

We take the state-space as $S = \{0, 1, 2, \ldots\}$. Under the assumptions of Theorem 1.1 of Billingsley (1961), it follows that there is a consistent solution of the likelihood equations which is asymptotically normal with mean vector 0 and the variance-covariance matrix $[I(\theta)]^{-1}$.

Example 2.2.7 Poisson Markov Sequence

A Poisson-Markov sequence $\{X_t, t = 0, 1, \ldots\}$ is defined as follows:

A1. $\{Y_t, \ t = 0, 1, \ldots\}$ is a sequence of i.i.d. Poisson random variables with $E(Y_0) = \lambda$.

A2. $P[Z_{t+1} = z | X_t = x, \ldots] = \binom{x}{z} p^z (1 - p)^{x-z}$

A3. Y_t is independent of $X_0, X_1, \ldots, X_{t-1}, Z_0, Z_1, \ldots, Z_t$, for each t.

A4. $X_{t+1} = Z_{t+1} + Y_{t+1}$.

In applications of a Poisson Markov sequence, X_t stands for the total number of members of the population at time t and Y_t are new recruits or newly joining members, whereas the random variable Z_t denotes survivors from the earlier day. Assumption A2 is equivalent to the assumption that an existing member survives for yet another time unit with probability p, irrespective of its age and independently of other members of the population. Assumption A3 says that Y_t, the number of new arrivals, is independent of the existing and past members (density-independent recruitment). This is, in fact, a discrete version of the $M|M|\infty$ queuing system. It follows that the one-step transition probability is given by

$$p_{ij} = P[X_{t+1} = j | X_t = i] = \sum_{z=0}^{\min(i,j)} \binom{i}{z} p^z (1-p)^{i-z} \frac{e^{-\lambda} \lambda^{j-z}}{(j-z)!}$$

We notice that $p_{ij} > 0 \ \forall \ (i, j)$, thus the Markov chain is irreducible and aperiodic. Further, p_{ij} is thrice differentiable in p and λ.

Now,

$$E(X_{t+1}|X_t) = E(Z_{t+1} + Y_{t+1}|X_t) = pX_t + \lambda.$$

If we assume that the process is stationary, $E(X_{t+1}) = E(X_t) = \mu$ (say), which implies that $\mu p + \lambda = \mu$ or $\mu = \lambda/(1 - p)$. A similar argument for variance implies that $\text{Var}(X_t)$ also equals μ for all t, if we assume stationarity. This suggests that the stationary distribution of the process is Poisson with mean μ. This is proved based on the following result, which is easy to prove.

Lemma 2.2.1 *If U has a Poisson distribution with mean λ, and if the distribution of V given $U = u$ is $Binomial(u, p)$ then V is Poisson with mean λp (if $U = 0$, we define $V = 0$).*

We recall the following well-known result for Markov chains.

Theorem 2.2.1 *A Markov chain $\{X_t, t = 0, 1, \ldots\}$ is strictly stationary, if and only if, X_0 and X_1 are identically distributed. Their common distribution is a stationary distribution .*

Proof Let $P[X_0 = j] = p_j, \ j \in S$. Then, $P[X_1 = j] = \sum_i P[X_0 = i, X_1 = j] = \sum_i p_i p_{ij}$. That is, $p_j = \sum_i p_i p_{ij} \ \forall j$, which satisfies the Definition 1.1.4 of a stationary distribution.

Now suppose that $X_0 \sim Poisson(\eta)$. Therefore, $X_1 \sim Poisson(\eta p + \lambda)$ by the Lemma and the Assumptions A1 and A2. Then X_0 and X_1 are identically distributed if and only f $\eta = \eta p + \lambda$. Thus, if $\eta = \lambda/(1 - p)$, the stationary distribution of the process is Poisson with mean $\mu = \lambda/(1 - p)$. Hence, the process is non-null persistent for all λ, p. We thus see that all the assumptions of a parametric model are satisfied.

Case I. Suppose $\{X_t, Y_t\}$, $t = 0, 1, \ldots, T$ are both observed. In this case, maximum likelihood estimation of parameters is very easy to carry out.

Case II. Now suppose that only X_t's have been observed. We re-parametrize the model in terms of μ and p. In this case, one can show that $\bar{X}$ is a good approximation to the MLE of μ (see Guttorp (1995), page 100). Estimation of p needs to be carried out by using iterative numerical procedures, such as Newton-Raphson. The Fisher Information matrix is rather involved and we may use the matrix $\hat{F}$ as its estimator.

2.3 Extensions of Markov Chain Models

Models based on the Logistic Regression.
Consider a two-state Markov chain. Let us write

$$\ln \frac{P[X_{t+1} = 1 | X_t = 0]}{P[X_{t+1} = 0 | X_t = 0]} = \beta_0 \text{ (base-line probability)}$$

and
$$\ln \frac{P[X_{t+1} = 1 | X_t = 1]}{P[X_{t+1} = 0 | X_t = 1]} = \beta_0 + \beta_1.$$

The t.p.m. can be written as

$$P = \begin{pmatrix} \frac{1}{1+e^{\beta_0}} & \frac{e^{\beta_0}}{1+e^{\beta_0}} \\ \frac{1}{1+e^{\beta_0+\beta_1}} & \frac{e^{\beta_0+\beta_1}}{1+e^{\beta_0+\beta_1}} \end{pmatrix}.$$

Though this appears to be only a re-parametrization, it serves to be useful and convenient. The above model in a compact form is written as

$$\ln \frac{P[X_{t+1} = 1 | X_t]}{P[X_{t+1} = 0 | X_t]} = \beta_0 + \beta_1 X_t.$$

The second order Markov chain with two states is modeled as

$$\ln \frac{P[X_{t+1} = 1 | X_t, X_{t-1}]}{P[X_{t+1} = 0 | X_t, X_{t-1}]} = \beta_0 + \beta_1 X_t + \beta_2 X_{t-1}.$$

The number of parameters in the above model is 3, whereas the saturated second order two-state Markov chain has 4 parameters. However, it must be pointed out that in such a model, unlike the saturated model, the transition probabilities are functions of the coding or numerical labels used to denote states of the chain. If the state-space refers to say linguistic classes, such as consonants and vowels, the above model need not be realistic.

An L order Markov Chain can be similarly defined by

$$\ln \frac{P[X_t = 1|X_{t-1}, X_{t-2}, \ldots, X_{t-L}]}{P[X_t = 0|X_{t-1}, X_{t-2}, \ldots, X_{t-L}]} = \beta_0 + \sum_{l=1}^{L} \beta_l X_{t-l}.$$

The above model has $L+1$ parameters as opposed to the saturated model which has 2^L parameters. A further advantage of such an approach is that we can incorporate time-dependent regressors also. Let $\{z_t\}$ be the sequence of values of regressors (possibly vector valued). Either the sequence $\{z_t\}$ is deterministic or if it is stochastic, the model is a conditional one. We write

$$\ln \frac{P[X_t = 1|X_{t-1}, X_{t-2}, \ldots, X_{t-L}, z_t]}{P[X_t = 0|X_{t-1}, X_{t-2}, \ldots, X_{t-L}, z_t]} = \beta_0 + \sum_{l=1}^{L} \beta_l X_{t-l} + \gamma' z_t.$$

A major advantage of this approach is that we can use any statistical package which analyzes logistic regression models.

M state L order Markov chain it based on Polytomous regression *model*. The logistic regression model for two categories can be extended to several categories, which is known as polytomous or multinomial (logistic) regression model. A polytomous regression model can be used to define an L order Markov chain with M states. Let $\tilde{P}[X_{L+1} = i] = P[X_{L+1} = i|X_1, X_2, \ldots, X_L]$, $i = 1, 2, \ldots, M$. Then,

$$\ln \frac{\tilde{P}[X_{L+1} = i]}{\tilde{P}[X_{L+1} = M]} = \beta_{0i} + \beta_{1i} X_1 + \cdots + \beta_{Li} X_L, \quad i = 1, 2, \ldots, M - 1.$$

$$\tilde{P}[X_{L+1} = M] = \frac{1}{1 + \sum_{i=1}^{M-1} \sum_{\ell=1}^{L} \exp\{\beta_{0i} + \beta_{\ell i} X_\ell\}}.$$

The above model has $(L + 1)(M - 1)$ parameters, far less than the corresponding saturated L-order model, which has $M^L(M - 1)$ parameters. Thus, such a model has the advantage of having less parameters and any software which analyzes the polytomous logistic regression data can be easily used to get the maximum likelihood estimators and estimators of their asymptotic variance-covariance matrix. Most of the packages include tests for significance of regression parameters. Such procedures can be used for testing of an order of a Markov chain . Analysis of higher order Markov chain models can also be carried out based on log-linear contingency table models, see Davison (2003).

Raftery's Mixture Transition Distribution Model

As observed earlier, a higher order M-state Markov chain model has too many parameters. An important higher order model with a significantly less number of parameters is due to Raftery (1985) and it is known as the Mixture Transition Distribution (MTD) model . The MTD model of order L is defined by

$$P[X_t = x_t | X_{t-1} = x_{t-1}, \ldots, X_{t-L} = x_{t-L}, \ldots, X_1 = x_1, X_0 = x_0]$$

$$= \sum_{l=1}^{L} \lambda_l p_{x_{t-l} x_t},$$

whenever the conditional probability on the left-hand side is defined. In the above, $P = ((p_{ij}))$ is an $M \times M$ stochastic matrix and $\{\lambda_l, l = 1, 2, \ldots, L\}$ constitutes a probability distribution. A probabilistic interpretation of the above model is as follows. Nature chooses the lag l with probability λ_l. If the chosen lag is r and if $X_{t-r} = i$, the conditional probability of $X_t = j$ given the chosen lag and the entire past is p_{ij}. The number of parameters of an MTD model of order L is $(L - 1) + M(M - 1)$, far less than the corresponding saturated model which has $M^L(M - 1)$ parameters. Adke and Deshmukh (1988) have shown that, if the matrix P is positively regular, i.e., if there exists a t such that all the elements of P^t are positive, the MTD model has the property that $P[X_{t+s} = j | X_s = i]$ converges to π_j, as $t \to \infty$ for every i, j, where π_j is, in fact, the unique stationary distribution of a Markov chain whose one-step t.p.m. is given by P. If X_0 follows the distribution π_j, the MTD model is stationary with π_j as the common distribution of X_t for all t. This result is useful in establishing properties of the MLE. Further, the MTD model is a parametric Markov model of order L and it can be shown to satisfy the Cramer regularity conditions . For a detailed discussion of MTD models including numerical procedures for estimation of parameters, we refer to Berchtold and Raftery (2002).

2.4 Hidden Markov Chains

Let $\{Y_t, t \geq 1\}$ be a stationary, irreducible, and aperiodic Markov chain on the state-space $\{1, 2, \ldots, M\}$ with P as the one-step t.p.m. and $\{\pi_i, i = 1, 2, \ldots, M\}$ as the unique stationary distribution. The chain $\{Y_t, t \geq 1\}$ is not observable. We observe the process $\{X_t, t \geq 1\}$, the state-space of which is $\{1, 2, \ldots, N\}$. The conditional distribution of X_t is given by

$$P[X_t = k | Y_t = j, Y_{t-1}, \ldots Y_1, X_{t-1}, X_{t-2}, \ldots, X_1] = P[X_t = k | Y_t = j] = q_{jk},$$

where $Q = ((q_{jk}))$ is an $M \times N$ matrix. The random sequence $\{X_t, t \geq 1\}$ on the state-space $\{1, 2, \ldots, N\}$ is known as a Hidden Markov chain. (In literature, $\{(X_t, Y_t) t \geq 1\}$ is sometimes described as a Hidden Markov chain.) In general, $\{X_t, t \geq 1\}$ does not satisfy Markov property. There are three important issues to be addressed.

(1) To derive likelihood function, i.e., $P[X_1 = x_1, X_2 = x_2, \ldots, X_T = x_T]$.

(2) To carry out state estimation, i.e., to derive the predictive distribution of the underlying chain :

$$P[Y_t = y_t | X_1 = x_1, \ldots, X_T = x_T], \ t = 1, 2, \ldots, T.$$

Prediction of Y_{T+j}, $j \geq 1$ may also be of interest.

(3) To carry out maximum likelihood estimation of the two unknown matrices P and Q

It is easy to see why (1) is involved: the likelihood function is a sum over T variables which correspond to the unobserved states of Y_t, $t = 1, 2, \ldots, T$. However, there exist recursive algorithms for (2) and (3) above, which are easy to implement.

Forward Algorithm. Define

$$\alpha_i(t) = P[X_1 = x_1, \ldots, X_{t-1} = x_{t-1}, Y_t = i], \ t = 2, 3, \ldots, T, \ i = 1, 2, \ldots, M,$$
$$\alpha_i(1) = P[Y_1 = i] = \pi_i.$$

The last equation defines the initialization of the algorithm. If any of the event $X_1 = x_1, \ldots, X_{t-1} = x_{t-1}$ is not well-defined, we take that event as Ω instead of ϕ, the empty set. In a recursive algorithm, we assume that $\alpha_i(1)$ are given for all i and find $\alpha_i(t)$ for $\forall \ i$ and $\forall \ t = 2, \cdots, T$. We further define

$$\alpha_i(T + 1) = P[X_1 = x_1, X_2 = x_2, \ldots, X_T = x_T, Y_{T+1} = i].$$

Then, it is easily seen that the likelihood function is given by

$$P[X_1 = x_1, X_2 = x_2, \ldots, X_T = x_T] = \sum_{i=1}^{M} \alpha_i(T + 1). \tag{2.3}$$

Next, by the defining properties of the two processes, we get

$$\alpha_i(2) = P[X_1 = x_1, Y_2 = i]$$
$$= \sum_{j=1}^{M} P[X_1 = x_1, Y_1 = j, Y_2 = i]$$
$$= \sum_{j=1}^{M} \alpha_j(1) q_{jx_1} p_{ji}.$$

In general,

$$\alpha_i(t + 1) = \sum_{j=1}^{M} \alpha_j(t) q_{jx_t} p_{ji}. \tag{2.4}$$

Next, we discuss the following Backward algorithm.

Backward Algorithm. Let

$$\beta_i(t) = P[X_t = x_t, \ldots, X_T = x_T | Y_t = i], \ t = 1, 2, \ldots, T, \ i = 1, 2, \ldots, M,$$

$$\beta_i(T + 1) = 1.$$

We note that

$$\beta_i(1) = P[X_1 = x_1, \ldots, X_T = x_T | Y_1 = i]$$
$$\beta_i(1)\pi_i = P[X_1 = x_1, \ldots, X_T = x_T, Y_1 = i],$$
$$\beta_i(T) = P[X_T = x_T | Y_T = i] = q_{i x_T}.$$

The Backward algorithm involves expressing $\beta_i(t)$ in term of $(\beta_i(t+1), \ldots, \beta_M(t+1))$. We observe that

$$\beta_i(t) = P[X_t = x_t, \ldots, X_T = x_T | Y_t = i]$$
$$= P[X_t = x_t, \ldots, X_T = x_T, Y_t = i]/\pi_i.$$

Then, from the properties of the Hidden Markov chain,

$$\beta_i(t) = \sum_{j=1}^{M} P[X_t = x_t, X_{t+1} = x_{t+1}, \ldots, X_T = x_T, Y_{t+1} = j | Y_t = i]$$

$$= \sum_{j=1}^{M} P[X_t = x_t, X_{t+1} = x_{t+1}, \ldots, X_T = x_T, Y_t = i, Y_{t+1} = j]/P[Y_t = i]$$

$$= \sum_{j=1}^{M} P[Y_t = i, X_t = x_t, Y_{t+1} = j, X_{t+1} = x_{t+1},$$
$$X_{t+2} = x_{t+2}, \ldots, X_T = x_T]/P[Y_t = i]$$

$$= \sum_{j=1}^{M} P[Y_t = i]P[X_t = x_t | Y_t = i]P[Y_{t+1} = j | X_t = x_t, Y_t = i]$$
$$P[X_{t+1} = x_{t+1}, \ldots, X_T = x_T | Y_{t+1} = j, X_t = x_t, Y_t = i]/P[Y_t = i]$$

$$= \sum_{j=1}^{M} q_{i x_t} p_{ij} \beta_{t+1}(j).$$

Let $\mathbf{X} = (X_1, X_2, \ldots, X_T)$ and $\mathbf{x} = (x_1, x_2, \ldots, x_T)$. Combining Backward and Forward Algorithms, we have

$$
\begin{aligned}
&P[X_1 = x_1, \ldots, X_t = x_t, X_{t+1} = x_{t+1}, \ldots, X_T = x_T, Y_t = i] \\
&= P[X_1 = x_1, \ldots, Y_t = i, X_t = x_t, X_{t+1} = x_{t+1}, \ldots, X_T = x_T] \\
&= P[X_1 = x_1, \ldots, X_{t-1} = x_{t-1}, Y_t = i] \\
&\quad \times P[X_t = x_t, X_{t+1} = x_{t+1}, \ldots, X_T = x_T | Y_t = i] \\
&= \alpha_i(t)\beta_i(t).
\end{aligned}
$$

Thus,

$$
P[\mathbf{X} = \mathbf{x}, Y_t = i] = \alpha_i(t)\beta_i(t)
$$

and therefore,

$$
P[\mathbf{X} = \mathbf{x}] = \sum_{i=1}^{M} \alpha_i(t)\beta_i(t).
$$

Now, if we had observed $Y_1, \ldots, Y_T$ also, the ML estimates of p_{ij} and q_{ik} would have been $\hat{p}_{ij} = \frac{\sum_t I[Y_t=i, Y_{t+1}=j]}{\sum_t I[Y_t=i]}$ (cf. Sect. 1.1) and $\hat{q}_{ik} = \frac{\sum_t I[Y_t=i, X_t=k]}{\sum_t I[Y_t=i]}$ respectively. The Baum-Welsch algorithm which we discuss below, computes conditional expectation of $I[Y_t = i, Y_{t+1} = j]$ and $I[Y_t = i, X_t = k]$ given the observations $\mathbf{X}$. We have

$$
\begin{aligned}
p_t(i, j) &= P[Y_t = i, Y_{t+1} = j | \mathbf{X} = \mathbf{x}] \\
&= \frac{P[Y_t = i, Y_{t+1} = j, \mathbf{X} = \mathbf{x}]}{P[\mathbf{X} = \mathbf{x}]}.
\end{aligned}
$$

Now, from the properties of a Hidden Markov chain,

$$
\begin{aligned}
&P[Y_t = i, Y_{t+1} = j, \mathbf{X} = \mathbf{x}] \\
&= P[X_1 = x_1, \ldots, X_{t-1} = x_{t-1}, Y_t = i, X_t = x_t, Y_{t+1} = j, \\
&\qquad X_{t+1} = x_{t+1}, \ldots, X_T = x_T] \\
&= \alpha_i(t)q_{ix_t} p_{ij} \beta_j(t+1).
\end{aligned}
$$

Hence, the likelihood is given by

$$
P[\mathbf{X} = \mathbf{x}] = \sum_{\ell=1}^{M} \sum_{k=1}^{M} \alpha_\ell(t)q_{\ell x_t} p_{\ell k} \beta_k(t+1).
$$

and

$$p_t(i, j) = \frac{\alpha_i(t) q_{i x_t} p_{ij} \beta_j(t+1)}{\sum_{\ell=1}^{M} \sum_{k=1}^{M} \alpha_\ell(t) q_{\ell x_t} p_{\ell k} \beta_k(t+1)}. \tag{2.5}$$

We may observe that the likelihood is also given by

$$L(P, Q) = P[X_1 = x_1, \dots, X_T = x_T] = \sum_{i=1}^{M} \beta_i(1) \pi_i. \tag{2.6}$$

We do not use the likelihood (2.3) or (2.6) for ML estimation, however, it is needed while comparing the Hidden Markov model with competing models. The recursive algorithm is as follows. The current estimate of p_{ij} is given by

$$\hat{p}_{ij} = \frac{\sum_t p_t(i, j)}{\sum_j \sum_t p_t(i, j)}.$$

Let

$$\gamma_i(t) = \sum_{j=1}^{M} p_t(i, j)$$

be the probability that the state i is observed at time t. Let $A(k) = \{t | I[X_t = k]\}$. The current estimate of q_{ik} is then given by

$$\hat{q}_{ik} = \frac{\sum_{t \in A(k)}^{N} \gamma_i(t)}{\sum_{t=1}^{N} \gamma_i(t)}. \tag{2.7}$$

We begin the algorithm with arbitrary estimates of the matrices P and Q and update their values as given above. The procedure continues until successive values differ by a pre-assigned tolerance. The algorithm is known as Baum-Welch or Forward-Backward algorithm. It is, in fact, one of the early versions of the EM algorithm, frequently used in the context of incomplete observations or samples with missing data.

For State Estimation, we need to obtain

$$\arg \max_{y_1,\dots,y_T} P[Y_1 = y_1, \dots, Y_T = y_T | X_1 = x_1, \dots, X_T = x_T].$$
$$= \arg \max_{y_1,\dots,y_T} P[Y_1 = y_1, \dots, Y_T = y_T, X_1 = x_1, \dots, X_T = x_T].$$

To do so, we apply the *Viterbi Algorithm* which is also recursive in nature. Let

$$\delta_j(t) = \max_{y_1, y_2, \dots, y_{t-1}} P[Y_1 = y_1, \dots, Y_{t-1} = y_{t-1}, Y_t = j, X_1 = x_1, \dots, X_{t-1} = x_{t-1}].$$

The initialization is carried out by

$$\delta_j(1) = \pi_j, \quad j = 1, 2, \dots, M,$$

the stationary distribution of the chain $\{Y_t, t \geq 1\}$. It can be shown that

$$\delta_j(t+1) = \max_{i=1,2,\dots,M} [\delta_i(t) p_{ij} q_{jx_t}].$$

Now, let

$$\psi_j(t+1) = \arg\max_{i=1,2,\dots,M} [\delta_i(t) p_{ij} q_{jx_t}].$$

The variable $\psi_j(t)$ records the "node of the incoming arc" that has resulted in this most probable path. The algorithm terminates with

$$\hat{Y}_{T+1} = \max_{i=1,2,\dots,M} \delta_i(T+1)$$

and

$$\hat{Y}_t = \psi_{\hat{Y}_{t+1}}(t+1),$$

which are predictors of $Y_1, Y_2, \dots, Y_T$. In the above algorithm, if there are ties, we break them randomly. Also, the algorithm assumes that the model parameters are known. In practice, it is implemented by replacing the unknown parameters (P, Q) by their MLEs.

The above discussion of the Backward-Forward algorithm is based on Manning and Schütze (1999). We refer to MacDonald and Zucchini (1997) for a thorough account of Hidden Markov processes on general state-spaces and their applications.

2.5 Aggregate Data from Finite Markov Chains

Model I. Here, we observe N i.i.d. finite Markov chains, each having the t.p.m. P. At each time unit $t = 1, 2, \dots, T$, we observe the number of Markov chains in the state i. However, transitions from a state i to a state j of these individual Markov chains are not available. Let $N(t, i)$ = Number of units or Markov chains in the state i at time t.

Notice that $\sum_i N(t, i) = N \; \forall t$. It is easy to see that

$$E[N(t, j) | N(t-1, 1), \dots, N(t-1, M)] = \sum_{i=1}^{M} N(t-1, i) p_{ij}.$$

This leads to a regression model, where the vector Y is the responses $N(t, j)$'s for $t = 2, 3, \dots, N$ and $j = 1, 2, \dots, M$. The regression vector β is the transition probabilities p_{ij}'s written in a conveniently chosen column form. Random variables

$N(t-1, i)$'s act as regressors to lead to a regression setup $E[Y] = X\beta$, in standard notations of regression analysis. We then have the Ordinary Least Square(OLS) estimator $\hat{\beta} = (X'X)^{-1}X'Y$. (One may take only the first $M - 1$ elements of each row of P and $N(t, i)$ for $i = 1, 2, \ldots, M - 1$). We note that variances of the response variables are not the same, also, they are not independent. We then need to consider the Weighted Least Squares (WLS). Both OLS and WLS estimators ignore the property that the transition probabilities are non-negative. (It is known that the OLS satisfies the condition that each of the row sum is 1, (cf. Lee et al. (1970), p. 34)) Thus, a better strategy is to minimize

$$(Y - X\beta)'(Y - X\beta)$$

subject to the constraints (i) $p_{ij} \geq 0$, $\forall (i, j)$ (ii) $\sum_{j=1}^{M} p_{ij} = 1 \,\forall\, i$. This is a constrained optimization problem (in fact, a Quadratic Programming Problem) and it can be shown to have a unique solution. A software package such as MATLAB or GAUSS can be used to get a solution to the optimization problem.

It can be shown that the sequence of $M \times 1$ random vectors $\{(N(t, 1), \ldots, N(t, M))$, $t \geq 1\}$ forms a Markov chain. Further, the conditional distribution of the random vector $(N(t, 1), N(t, 2), \ldots, N(t, M))$ is a multinomial distribution with parameters N and the cell probabilities $\sum_i (N(t-1, i)/N)p_{i1}, \sum_i (N(t-1, i)/N)p_{i2}, \ldots, \sum_i (N(t-1, i)/N)p_{iM}$. Verification of the regularity conditions is straightforward.

Model II. It is not necessary that we observe the same N individuals throughout. Thus, at each time, we observe $N(t, i)$, $i = 1, 2, \ldots, M$ and $\sum_i N(t, i) = N(t)$ which need not be the same for all t. Technically, the T random vectors

$$\{N(1, 1), \ldots, N(1, M)\}, \{N(2, 1), \ldots, N(2, M)\}, \ldots, \{N(T, 1), \ldots, N(T, M)\}$$

are independently distributed. Now,

$$E[N(t + 1, j)] = \sum_i \pi_{t,i} p_{ij},$$

where $\pi_{t,i}$ is the probability that a randomly chosen person (at time t) is in the state i. This is a case of moment estimation, where we first estimate $\pi_{t,i}$ by the observed proportions $N(t, i)/N(t)$. Replacing this estimate in the above, we again get the situation similar to Model I and employ the LSEs. Lee et al. (1970) have an extensive review of the various methods to estimate the transition probabilities.

In each of the above cases, we can allow either T to tend to ∞ or $N(t)$ to tend to ∞ for each t (or both). In either case, it can be shown that LSE/MLE is consistent and asymptotically normal with appropriate norming. When the process reaches equilibrium, for large N, the relative frequencies of M states at time t are close to the unique stationary distribution and therefore to each other. Since they act as regressors, the matrix X of the above regression model turns out to be nearly singular. This leads to a multi-collinearity problem. Inderdeep Kaur and Rajarshi (2012) discuss ridge-regression estimators which offer a considerable improvement over the LSE in terms of the total mean squared error.

Books by Guttorp (1995), Davison (2003) (Chap. 6) and Lindsey (2004) are good sources of inference in Markov chains and related topics.

References

Adke, S.R., Deshmukh, S.R.: Limit distribution of a higher order Markov chains. J. Roy. Stat. Soc. Ser. B **50**, 105–108 (1988)

Berchtold, A., Raftery, A.E.: Mixture transition distribution model for high-order markov chains and non-gaussian time series. Stat. Sci. **17**, 328–356 (2002)

Billingsley, P.: Statistical Inference for Markov Processes. The University of Chicago Press, Chicago (1961)

Cappé, O., Moulines, E., Ryden, T.: Inference in Hidden Markov Models. Springer, New York (2005)

Davison, A.C.: Statistical Models. Cambridge University Press, Cambridge (2003)

Elliot, R.J., Aggaoun, L., Moore, J.B.: Hidden Markov Models: Estimation and Control. Springer, New York (1995)

Guttorp, P.: Stochastic Modeling of Scientific Data. Chapman and Hall, London (1995)

Inderdeep Kaur, Rajarshi: Ridge regression for estimation of transition probabilities from aggregate data. Commun. Stat. Simul. Comput. **41**, 524–530 (2012)

Katz, R.W.: On some criteria for estimating the order of a markov chain. Technometrics **23**, 243–249 (1981)

Lee, T.C., Judge, G.G., Zellner, A.: Estimating the Parameters of the Markov Probability Model from Aggregate Time Series Data. North Holland, Amsterdam (1970)

Lindsey, J.K.: Statistical Analysis of Stochastic Processes in Time. Cambridge University Press, Cambridge (2004)

MacDonald, I., Zucchini. W.: Hidden Markov and Other Models for Discrete-Valued Time Series. Chapman and Hall, London (1997)

Manning, C.D., Schütze, H.: Statistical Natural Language Processing. MIT Press, Cambridge (1999)

Raftery, A.E.: A model for high order Markov chains. J. Roy. Stat. Soc. Ser. B **47**, 528–539 (1985)

Chapter 3
Non-Gaussian ARMA Models

Abstract We discuss stationary AR and ARMA time series models for sequences of integer-valued random variables and continuous random variables. Stationary distribution of these models is non-Gaussian. Such models can be broadly described as extensions of Gaussian ARMA models, which have been very widely discussed in the time series literature. These non-Gaussian AR models share two important properties with a linear AR(1) model: (i) the conditional expectation of X_t is a linear function of the past observation and (ii) the auto-correlation function (ACF) has an exponential decay. However, the conditional variance of an observation is frequently a function of the past observations. These models are formed so as to have a specific form of the stationary distribution. Stationary distributions include standard discrete distributions such as binomial, geometric, Poisson, and continuous distributions such as exponential, Weibull, gamma, inverse Gaussian, and Cauchy. In some cases, maximum likelihood estimation is tractable. In other cases, regularity conditions are not met. Estimation is then carried out based on properties of the marginal distribution of the process and mixing properties such as strong or ϕ-mixing are useful to derive properties of the estimators.

3.1 Integer Valued Non-Negative Auto-Regressive Models

We begin with the definition of a thinning operator.

Definition 3.1.1 Let $\rho \in [0, 1)$ and let X be a non-negative integer valued random variable. Then $\rho \circ X = Binomial(X, \rho)$ defines a thinning operator $\rho \circ X$. If $X = 0, \rho \circ X - 0$.

We notice that $\rho \circ X$ defines a conditionally Binomial random variable. The Integer Non-negative AR(1) (INAR(1)) sequence, introduced by Al-Osh and Alzaid (1987) is defined as follows.

M. B. Rajarshi, *Statistical Inference for Discrete Time Stochastic Processes*,
SpringerBriefs in Statistics, DOI: 10.1007/978-81-322-0763-4_3,
© The Author(s) 2012

Definition 3.1.2 INAR(1) Sequence. Let $\{\varepsilon_t, t \geq 1\}$ be a sequence of i.i.d. non-negative integer valued random variables, assumed to be independent of X_0. The INAR(1) sequence is defined by

$$X_t = \rho \circ X_{t-1} + \varepsilon_t, \quad t \geq 1.$$

We write $\bar{\rho} = 1 - \rho$, $\lambda = E(\varepsilon_1)$ and $\eta^2 = \text{Var}(\varepsilon_1)$, which is assumed to be finite. The conditional mean and conditional variance of X_t given the past are given by

$$E[X_t|X_{t-1}] = \rho X_{t-1} + \lambda, \quad \text{Var}(X_t|X_{t-1}) = \rho\bar{\rho}X_{t-1} + \eta^2.$$

The INAR(1) sequence is Markovian and it is stationary, if $Y = \sum_s \rho^s \circ \varepsilon_s$ is a proper random variable and if X_0 has the same distribution as that of Y. The INAR(1) sequence is weakly stationary with

$$E(X_t) = \frac{\lambda}{\bar{\rho}}, \quad \text{Var}(X_t) = \frac{\rho\lambda + \eta^2}{1 - \rho^2}.$$

It can be shown that the ACF of the sequence is $\{\rho^t, t \geq 0\}$. The sequence does not admit negative correlations of any lag.

Example 3.1.1 Poisson Markov sequence (Example 2.2.7)
The Poisson Markov sequence is INAR(1) sequence with the distribution of ε as *Poisson*(λ).

Example 3.1.2 Geometric AR sequence
Let $\{I_t, t \geq 1\}$ and $\{E_t, t \geq 1\}$ be independent sequences of random variables. Each of these sequence is a sequence of non-negative integer valued i.i.d. random variables. Let I_t be a Bernoulli random variable with $E(I_t) = \bar{\rho}$ and let E_t be a geometric random variable with parameter θ, the probability of "success". Then, the INAR(1) sequence has a geometric stationary distribution if and only if ε_t has a distribution which is the same as that of $I_t E_t$. The first two conditional and unconditional moments can be simplified in terms of those of a geometric random variable. Let $\bar{\theta} = 1 - \theta$. The one-step transition probability of the Geometric INAR(1) sequence is given by

$$p_{xy} = \begin{cases} \bar{\rho}\bar{\theta} \sum_{k=0}^{y-1} \binom{x}{k}\rho^k(\bar{\rho}\theta)^{x-k} + \binom{x}{y}\rho^{y+1}(\bar{\rho})^{x-y}, & y \leq x \\ \bar{\rho}\bar{\theta}\theta^{y-x}(\rho + \bar{\rho}\theta)^x, & y > x. \end{cases}$$

The above process was introduced and studied by McKenzie (1985, 1986).

In general, to define a stationary process with the p.g.f. $P_X(s)$ of the stationary distribution, $P_\varepsilon(s)$, the p.g.f. of ε_t, needs to satisfy the identity

$$P_\varepsilon(s) = \frac{P_X(s)}{P_X(1 - \rho + \rho s)}.$$

Example 3.1.3 Negative Binomial AR sequence
An INAR(1) sequence with a Negative Binomial distribution as the stationary distribution is defined by

$$P_\varepsilon(s) = \left(\frac{\lambda + \rho s}{\lambda + s}\right)^\beta,$$

for $\lambda > 0$, $\beta > 0$. The p.g.f. of the marginal distribution of X_1 is given by $(\lambda/(\lambda + \rho s))^\beta$. The transition probabilities can be obtained in terms of the Binomial probabilities and the probability distribution of ε.
A detailed analysis of the above examples can be found in McKenzie (1986). A variation of the above type of AR models is described in the following example.

Example 3.1.4 Binomial Markov sequence
A Binomial Markov sequence is defined by

$$X_0 \sim \text{Binomial}(N, p)$$

$$X_t = \rho \circ X_{t-1} + \beta \circ (N - X_{t-1}), \ t \geq 1,$$

where $\beta = (1-\rho)p/(1-p)$, if $\beta \leq 1$, cf. McKenzie (1985). If $\beta > 1$, we interchange the role of p and ρ. It is easy to see that the stationary distribution of the sequence is *Binomial*(N, p). This Markov chain corresponds to the grouped data with the two states 0 and 1 and N i.i.d. two state Markov chains. The random variable X_t refers to the number of individuals in state 1 at time t. The one-step t.p.m. can be found easily.
The above models are Markovian of order one and in each case, the transition probabilities are easily found. Regularity conditions discussed in the Chap. 2 are also satisfied and thus we carry out ML estimation.

3.2 Auto-Regressive Models for Continuous Random Variables

Let us consider an auto-regressive equation

$$X_t = \rho X_{t-1} + \varepsilon_t,$$

for $t \geq 1$. It is assumed that both X_t and ε_t are non-negative random variables. We specify a continuous probability distribution for the stationary distribution of the above sequence and we seek whether there exists an appropriate innovation distribution. Let $\phi_\varepsilon(s)$ and $\phi_X(s)$ be the Laplace transforms of ε_t and X_t respectively. Under the assumption of independence of ε_t and X_{t-1}, we have

$$\phi_X(s) = \phi_X(\rho s)\phi_\varepsilon(s). \tag{3.1}$$

A stationary sequence exists, if the unique solution of the above equation in $\phi_\varepsilon(s)$ is a proper Laplace transform.

Example 1 Inverse Gaussian AR sequence
Here, we define an Inverse Gaussian p.d.f. by

$$f(x, \mu, \lambda) = \left(\frac{\lambda}{2\pi x^3}\right)^{1/2} \exp\left(\frac{-\lambda(x-\mu)^2}{2\mu^2 x}\right), \quad x > 0; \ \mu > 0, \lambda > 0.$$

The Laplace transform of the above distribution is given by

$$\phi_X(s) = \exp\left\{\frac{\lambda}{\mu}\left(1 - \left(1 + \frac{2\mu^2 s}{\lambda}\right)^{1/2}\right)\right\}.$$

Pillai and Satheesh (1992) have proved that in this case, $\phi_\varepsilon(s) = \phi_X(s)/\phi_X(\rho s)$ is the Laplace transform of a proper distribution for every $\rho \in [0, 1]$. The Laplace transform of the corresponding innovation distribution is given by

$$\phi_\varepsilon(s) = \exp\left[-\frac{\lambda}{\mu}\left\{\left(1 + \frac{2\mu^2}{\lambda}\right)^{1/2} - \left(1 + \frac{2\rho\mu^2}{\lambda}\right)^{1/2}\right\}\right].$$

There is no closed form for the p.d.f. which is obtained by inverting the above Laplace transform. Abraham and Balakrishna (1999) discuss the special case when $\mu \to \infty$ in the above. In this special case, the Laplace transform $\phi_\varepsilon(s)$ is given by $\exp\{-\sqrt{(2\lambda s)}\}$ and the corresponding transition p.d.f. is given by

$$f(y|x) = \begin{cases} \left[\frac{\lambda(1-\sqrt{\rho})^2}{2\pi(y-\rho x)^3}\right]^{1/2} \exp\left[-\frac{\lambda(1-\sqrt{\rho})^2}{2(y-\rho x)}\right], & y \geq \rho x \\ 0, & \text{otherwise.} \end{cases} \tag{3.2}$$

The range of X_{t+1} depends upon both ρ and X_t and the regularity assumptions are not met. However, non-negativity of innovations suggests the following estimator

$$\hat{\rho} = \min_{1 \leq t \leq T} \frac{X_t}{X_{t-1}},$$

which has been studied by Feigin and Resnick (1992). Their main theorem is as follows.

Theorem 3.2.1 *(Feigin and Resnick 1992). Let $\{X_t, t \geq 1\}$ be a stationary process defined by $X_t = \rho X_{t-1} + \varepsilon_t$, where $\rho \in [0, 1)$ and $\{\varepsilon_t,\ t \geq 1\}$ is a sequence of i.i.d. non-negative random variables. Let G be the d.f. of ε_1. Assume that the followings hold.*
A1. For some $\eta > 0$,

$$\lim_{s \to \infty} \frac{1 - G(sx)}{1 - G(s)} = x^{-\eta}$$

A2.

$$E\left[\varepsilon_1^{-\beta}\right] < \infty \ for\ some\ \beta > \eta.$$

Then $P[b(T)(\hat{\rho} - \rho) > x] \to \exp[cx^{-\eta}]$ as $T \to \infty$, where

$$b(T) = [1/(1 - G)] * (T), \qquad (H * (T) = \inf\{x | H(x) \geq T\}),$$

and c is given by

$$c = \int_0^\infty \left[1 - \prod_{n-0}^\infty \left(1 - G(\rho^n s)\right) \right] \eta s^{-\eta - 1} ds.$$

Abraham and Balakrishna (1999) point out that for the model (3.2) the assumption A1 is satisfied with $\eta = 1/2$. Moreover, since $1/\varepsilon_t$ has a gamma distribution, all the moments are finite and A2 is satisfied. However, it is difficult to identify the limiting distribution. They, therefore, develop an estimator of λ which is based on the empirical Laplace transform of the marginal distribution of X_1. Its CAN property is established by applying mixing properties of the sequence and the Theorem 1.3.4. We can develop models based on the characteristic functions instead of Laplace transforms. If $|\rho| < 1$ in and if we allow $\phi(\tau)$ to denote the characteristic function instead of the Laplace transform in (3.1) $\phi_X(\tau) = \phi_X(\rho\tau)\phi_\varepsilon(\tau)$.

Example Cauchy Auto-regressive sequence
With $\phi_X(\tau) = \exp(-\delta \mid \tau \mid)$, we have $\phi_\varepsilon(\tau) = \exp\left(-\delta(1 - \mid \rho \mid) \mid \tau \mid\right)$. Thus, innovations also have a Cauchy distribution. The transition density is also Cauchy with location ρX_{t-1} and scale $\delta(1 - \mid \rho \mid)$. However, the likelihood analysis seems to be cumbersome. It is also not clear whether various regularity conditions are met. Applying the results of Athreya and Pantula (1986a,b); Balakrishna and Nampoothiri (2003) show that the Cauchy AR sequence is GSM . They suggest $\sqrt{T}$-CAN estimator of δ based on the fact that $P[\mid X_1 \mid \leq x] = (2/\pi)\tan^{-1}(x/\delta)$ and the strong mixing properties. Further, applying the general results regarding $\hat{\rho} = \sum_t X_t X_{t-1} / \sum_t X_{t-1}^2$, the sample auto-correlation of lag 1, it is shown that $\hat{\rho} \to \rho$ in probability and that as $T \to \infty$, $\sqrt{T/\ln T}(\hat{\rho} - \rho)$ has a non-trivial limiting distribution (cf. Brockwell and Davis 1987, p. 482).

3.3 Processes Obtained by Minification

In this section, we discuss first order stationary Markov sequences obtained by mini-fication . These have been introduced by (Tavares, 1980a), also see Gaver and Lewis (1980), Lewis and McKenzie (1991). Such a sequence is defined by

$$
X_t = \begin{cases} X_0, & t = 0 \\ \kappa \; \min\{X_{t-1}, \varepsilon_t\}, & t \geq 1, \; \kappa > 1 \end{cases}
$$

where $\{\varepsilon_t, t \geq 1\}$ is a sequence of i.i.d. non-negative random variables and X_0 is independent of $\{\varepsilon_t, t \geq 1\}$. Let $\overline{F}(x) = P(X_0 > x)$ and let $\overline{G}(x) = P(\varepsilon_1 > x)$. Then, the above sequence is stationary if and only if,

$$
\overline{G}(x) = \frac{\overline{F}(\kappa x)}{\overline{F}(x)}, \quad x \geq 0.
$$

Arnold and Hallett (1989) have shown that if the distribution of X_0 is chosen as

$$
\overline{F}(x) = \prod_{t=0}^{\infty} \overline{G}(x/\kappa^t), \tag{3.3}
$$

then the above system defines a stationary sequence with $\overline{F}(x)$ as the survival function of the common marginal distribution (this assumes that the above product is well defined and positive) . Let us define

$$
X_0 = \inf_{0 \leq t < \infty} \kappa^t \varepsilon_{-t}.
$$

Then, the survival function of X_0 is given by $\overline{F}(x)$ given by (3.3).
Let $\rho = 1/\kappa$ and $V_t = X_t/X_{t-1}, t \geq 1$. Adke and Balakrishna (1992) introduce the stopping time τ defined by

$$
\tau = \inf \{t \mid V_t = V_s \text{ for some } s < t\}.
$$

It follows that $X_\tau = \rho$ a.s.. Adke and Balakrishna (1992) show that the distribution of τ is Binomial$(2, \rho)$. This leads to a sequential sampling scheme under which the parameter ρ is estimated without any sampling error and thus, in a large sample analysis, one may assume that it is known. Adke and Balakrishna (1992) point out that such a result holds for the exponential AR (1) model and the gamma AR (1) model models in Gaver and Lewis (1980). Adke and Balakrishna (1992) derive ML and BLUE estimators of the mean of the process assuming that ρ is known. They also discuss two stage estimation procedures for estimation of parameters.

Based on Lewis and McKenzie (1991), Balakrishna and Jacob (2003) establish that the above minification sequence is ergodic and ϕ-mixing with the mixing coefficients given by $\phi(t) = \int_0^\infty \overline{F}(\kappa^t x)/\overline{F}(x) dF(x)$. Further, they show that the estimator

$$\hat{\kappa} = \max_t \frac{X_t}{X_{t-1}}$$

converges to κ a.s. Its limiting distribution has been studied and it has been shown that the limiting distribution is not Gaussian. One may carry out statistical inference for other parameters by taking $\hat{\kappa}$ as a known value of κ. Balakrishna and Jacob (2003) study the following exponential minification process in details. With $\overline{G}(x) = \exp(-(\kappa - 1)x/\mu)$, we have $\overline{F}(x) = \exp(-x/\mu)$, so that the stationary distribution is exponential with mean μ.

3.4 Product AR Models

Let $\{X_t, t \geq 1\}$ be a random sequence defined recursively by

$$X_t = X_{t-1}^\rho V_t, \quad 0 \leq \rho < 1,$$

where X_t and V_t are positive random variables for each t. McKenzie (1982) discusses the above model mainly for gamma random variables. One specifies a marginal distribution as the stationary distribution of the above sequence and investigates existence and form of distribution of V_t. Such processes are clearly Markovian, if V_t is a sequence of i.i.d. positive random variables. We denote these models by PAR(1). The conditional mean and variance are respectively given by

$$E(X_t \mid X_{t-1}) = \mu_V X_{t-1}^\rho, \quad \mathrm{Var}(X_t \mid X_{t-1}) = \sigma_V^2 X_{t-1}^{2\rho},$$

where μ_V and σ_V^2 respectively denote the mean and the variance of the random variable V_1. An advantage of these models is that the conditional distribution of X_t given X_{t-1} is absolutely continuous, unlike the exponential, Weibull and gamma models, discussed earlier. The exponential PAR(1) model is a special case. The simplest model is the lognormal model: the random variable $\ln(V_1)$ has $N\left((1 - \rho)\mu, (1 - \rho^2)\sigma^2\right)$, $|\rho| < 1$ distribution. The stationary distribution is lognormal with parameters μ and σ^2.

A Weibull model

Suppose that the innovation V has the distribution of $(\lambda/Y)^{\rho/\theta}$, where the positive random variable Y has a stable distribution with the Laplace transform $\exp(-\lambda s^\rho)$, $(\theta, \lambda > 0)$. It can be then shown that the survival function of the stationary distribution is given by $P[X > x] = \exp(-\lambda x^\theta)$ so that X has a Weibull distribution. We refer to Balakrishna and Shiji (2010) for likelihood analysis of the Weibull model.

An interesting case is $\theta = 1$, $\rho = 1/2$, in which case the distribution of V turns out to be a truncated Normal distribution $N(0, 4/\lambda)$, truncated from above at 0.

Balakrishna and Lawrence (2012) give a review of PAR(1) models.

3.5 More General Non-Gaussian Sequences

1. *Integer valued Non-negative Auto-regressive Moving Average models*
Let $\{\varepsilon_t, t \geq (1 - q)\}$ be a sequence of i.i.d. non-negative integer valued random variables. An Integer valued Moving Average model of order q is defined by

$$X_t = \varepsilon_t + \sum_{i=1}^{q} \beta_i \circ \varepsilon_{t+1-i} \quad t \geq 1.$$

Instead of Binomial thinning operator, we can also define a hyper-geometric thinning operator and define processes in terms of such an operator.
We now discuss INARMA (1, q) models. Let $\{Y_t, t \geq 1\}$ be an INAR(1) process which with $\{\varepsilon_t, t \geq (1 - q)\}$ as the corresponding ε-sequence, i.e.,

$$Y_t = \alpha \circ Y_{t-1} + \varepsilon_t.$$

Then, an INARMA (1, q) sequence is defined by

$$X_t = Y_{t-q} + \sum_{i=1}^{q} \beta_i \circ \varepsilon_{t+1-i}.$$

The ACF of the process is the same as that of a standard linear ARMA(1, q) process. Geometric and Negative binomial ARMA(1, q) sequences are defined by having β_i' as random variables.
An ARMA(p, 1) process is defined by

$$X_t = \sum_{i=1}^{p} \alpha_i \circ X_{t-i} + \varepsilon_t$$

cf. Du and Li(1991). It is assumed in such a definition that the successive binomial (thinning) experiments are independent and that innovations ε_t's are independent of thinning experiments. Further, $\sum_{i=1}^{p} \alpha_i < 1$. McCabe et al. (2011) assume that ε_t follows the distribution G, the functional form of which is not known. Likelihood of the process is easily written in terms of the parameters α_i's g_r's, where $g_r = P[\varepsilon_t = r]$. Observations restrict the range of r's to $\max\{0, \min_{t=p+1,\cdots,T}(x_t - \sum_{i=1}^{p} x_{t-i})\} \leq r \leq \max_{t=p+1,\cdots,T}(x_t)$. This leads to a non-parametric MLE (NPMLE). If $p + 4$th moment of G is finite and if $g_0 > 0$, the NPMLE is consistent and asymptotically normal, cf. Drost et al. (2009) and McCabe et al. (2011). In McCabe et al. (2011)(Sect. 2.2

of the paper), the predictive distribution of future observations is shown to be Fréchet differentiable as a function of α_i's and G. This is further used to prove the asymptotic normal distribution of the estimator of the predictive distribution.

2. *DARMA models*

These models have been introduced and extensively discussed by Jacobs and Lewis (1978 a,b, 1983), c. Let $\{Y_t, t \geq 1\}$ be a sequence of discrete i.i.d. random variables with $P[Y_1 = i] = \pi_i$. Let $\{U_t, t \geq 1\}$ and $\{V_t, t \geq 1\}$ be independent sequences of i.i.d. Bernoulli random variables with parameters $\beta \in [0, 1]$ and $\rho \in (0, 1)$ respectively. Let $\{D_t, t \geq 1\}$ and $\{A_t, t \geq 1\}$ be independent sequences of i.i.d. random variables with

$$P[D_t = r] = \delta_r, \quad r = 0, 1, 2, \ldots, N,$$

$$P[A_t = r] = \alpha_r, \quad r = 1, 2, \ldots, M,$$

where N is a non-negative integer and M is a positive integer. The DARMA $(M, N + 1)$ sequence $\{X_t, t \geq 1\}$ is then defined by

$$Z_t = V_t Z_{t-A_t} + (1 - V_t)Y_t, \quad t = -N, -N + 1, \ldots$$

$$X_t = U_t Y_{t-D_t} + (1 - U_t)Z_{t-(N+1)}, \quad t = 1, 2, \ldots.$$

The sequence $\{Z_t, t \geq -N\}$ is known as a DAR(p) sequence. We note that one needs to define a joint distribution of $Z_{-N-p}, \ldots, Z_{-N-1}$. It can be shown that there exists a stationary distribution for the process $\{Z_t, t \geq -N\}$. Further, if the initial distribution of the process $\{Z_t, t \geq -N\}$ is the stationary distribution, the process $\{X_t, t \geq 1\}$ is strictly stationary and its one-dimensional marginal distribution is given by π_i. It has been shown that the DARMA $(M, N + 1)$ process is ϕ-mixing. Methods of estimation have been mostly based on equating first few sample autocorrelations with those of the population auto-correlations. It needs to be stated that the process admits only non-negative auto-correlations. Special cases such as DAR(p), DARMA(1,1) are of interest and have auto-correlation structures similar to those of corresponding Box-Jenkins linear ARMA models. McKenzie (2003) reviews integer valued sequences of dependent variables.

3. *Generalized ARMA models*

The Generalized ARMA models have been studied in Benjamin et al. (2003). We assume that the conditional p.d.f. or p.m.f. of X_t given the past observations is given by

$$f(x_t \mid x_1, x_2, \ldots, x_{t-1}) = \exp\left(\frac{x_t v_t - b(v_t)}{\varphi} + d(x_t, \varphi)\right),$$

where v_t and φ are canonical parameters and $b(\cdot)$ and $d(\cdot)$ functions which define an exponential family of distributions. It follows from the properties of the exponential family that $\mu_t = E(X_t|X_1, X_2, \ldots, X_{t-1}) = b'(v_t)$ and that $\text{Var}(X_t|X_1, X_2, \ldots, X_{t-1}) = \varphi v(\mu_t) = \varphi b''(v_t)$. A Generalized ARMA model is given by

$$g(\mu_t) = \mu + \sum_{j=1}^{p} \phi_j (g(X_{t-j}) - \mu) + \sum_{j=1}^{q} \theta_j (g(X_{t-j}) - \mu_{t-j}),$$

for an appropriate choice of the function g. Here, ϕ_j's and θ_j's are respectively known as the auto-regressive and moving average parameters and μ is a real parameter.

Example 3.5.1 Generalized Poisson ARMA(p,q) model
Here the conditional distribution of X_t is Poisson and

$$\log(\mu_t) = \mu + \sum_{j=1}^{p} \phi_j (\log(X_{t-j}^*) - \mu) + \sum_{j=1}^{q} \theta_j (\log(X_{t-j}^*) - \mu_{t-j}),$$

where $X_s^* = \max(X_s, c)$ for some c, $0 < c < 1$.

Example 3.5.2 Binomial Logistic ARMA(p,q) model
The conditional distribution of X_t is $Binomial(N_t, \mu_t)$ and g is the logit function given by $g(u_t) = u_t/(N_t - u_t)$. As in the case of the Poisson model above, we need to introduce a threshold parameter and take $X_t^* = \min\left(\max(X_t, c), N_t - c\right)$ in the definition of link function.

Benjamin et al. (2003) also discuss the Gamma ARMA(p, q) model.

The above class of models is very flexible and includes a large number of sequences studied earlier in the literature. Conditions for stability of marginal mean and variance, for the identity link function, are as follows (see Appendix of Benjamin et al. (2003)).

(i) $E(X_t) = \mu$ for all t.
(ii) Let $\Psi(B) = [\Phi(B)]^{-1}\Theta(B)$ where $\Phi(B) = 1 - \phi_1 B - \phi_2 B^2 - \cdots - \phi_p B^p$ and $\Theta(B) = 1 + \theta_1 B + \theta_2 B^2 + \cdots + \theta_q B^q$ are the usual AR and MA polynomials associated with the AR and MA parameters respectively. The operator B is defined by $B(X_t) = X_{t-1}$. It is assumed that $\Phi(B)$ is invertible. Let $\Psi^{(2)}(B) = 1 + \psi_1^2 B + \psi_2^2 B^2 + \cdots$. Then, the common marginal variance of X_t is given by

$$\text{Var}(X_t) = \varphi E\left[\Psi(B)^{(2)}v(\mu_t)\right].$$

For example, for the variance stationary Poisson model, the common variance is given by $\mu(1+\sum_{t=1}^{\infty} \psi_t^2)$. Parameter space is constrained by positivity of conditional variances.
A limitation is that, except in the case when g is linear, conditions for stationarity and invertibility are difficult to investigate (in the linear case, the conditions are the same as those of a standard linear ARMA model). Benjamin et al. (2003) describe how simulation techniques can be used to determine constraint on the parameters to

have stationarity. They suggest to simulate time series from a Generalized ARMA model of various lengths such as 50, 100, 150, 200 and compute mean, variance, skewness and kurtosis at each stage. Stability of these statistics strongly indicates stationarity. The stationary distribution of a Generalized ARMA sequence is hard to derive and here again, one may employ simulation techniques.

4. *Generalization of Raftery's MTD model*

We assume that $\{X_t, t \geq 1\}$ is an L-order Markov sequence. Let $F(y|x_1, x_2, \ldots, x_L)$ be the distribution function of X_{L+1} given that $X_1 = x_1, X_2 = x_2, \ldots, X_L = x_L$. Then, a MTD model is given by

$$F(y|x_1, x_2, \ldots, x_L) = \sum_{r=1}^{L} \lambda_r G_r(y|x_{L-r}),$$

where $G_r(y|x_{L-r})$ is a d.f. in y. In practice, to have a model with less number of parameters, we take $G_r(y|x) = G(y|x)$. This model has an ACF similar to that of $AR(L)$ model. Berchtold and Raftery (2002) also modify the MTD model so that one can have $\{X_t, t \geq 1\}$ to be an i.i.d. sequence. The stationary distribution of the sequence is that of the stationary distribution of a first-order Markov sequence with $G(y|x)$ as the transition d.f. and under conditions similar to finite Markov chains, one can show that the sequence is ergodic and the ACF decays geometrically fast. Raftery and Tavare (1994) give estimation procedures and reference to the relevant software also.

Weiß (2009) defines an auto-regressive model of order L for a binomial process as follows. Let $\pi \in (0, 1)$ and $\rho \in [\max\{-\pi/(1 - \pi), -(1 - \pi)/\pi\}, 1]$. Let, further, $\beta = \pi(1 - \rho)$ and $\alpha = \beta + \rho$. Then, given the last L random variables, X_t equals $\alpha \circ X_{t-i} + \beta \circ (N - X_{t-i})$ with probability $\phi_i, i = 1, 2, \ldots, L, \sum_{i=1}^{L} \phi_i = 1$. It is assumed that events corresponding to all thinning operations are independent. The stationary distribution is $B(N, \pi)$.

5. *Generating non-Gaussian models from a Linear Gaussian model*

Here, we discuss a methodology to define a general non-Gaussian sequence obtained via a Gaussian sequence. This discussion is based on Block et al. (1990). Let $Y_t = \sum_{s=0}^{\infty} \psi_s \varepsilon_{t-s}$ and let $\mathcal{Y}$ be the corresponding linear, stationary, invertible Gaussian time series. We assume that $E(Y_t) = 0$ and $E(Y_t^2) = 1$. Let ϕ be the distribution function of Y_t. Let H be a continuous distribution function and let H^{-1} be its right inverse defined by

$$H^{-1}(p) = \begin{cases} \inf\{x \mid H(x) > p\}, & p < 1 \\ \sup\{x \mid H(x) < 1\}, & p = 1. \end{cases}$$

Then, the stochastic process $\{X_t\}$ is defined by

$$X_t = H^{-1}\{\phi(Y_t)\}.$$

Block et al. (1990) suggest the following approach, when H is an unknown distribution function. It is assumed that the Gaussian sequence $\mathcal{Y}$ is α-mixing (Example 1.3.5) and that conditions of Theorem 1.4.1 are satisfied. It follows that $\{X_t\}$ is also a strong mixing sequence. Let H_T be the empirical distribution function of $X_1, X_2, \ldots, X_T$. To estimate θ, consider $\tilde{Y}_t$ defined by

$$\tilde{Y}_t = \phi^{-1}[H_T(X_t)], \ t = 1, 2, \ldots, T.$$

We now regard $\tilde{Y}_t$'s as observations from the Gaussian sequence $\{Y_t, t \geq 1\}$ and compute the ML estimates. If the process $\{Y_t, t \geq 1\}$ is ARMA, any standard package which carries out statistical analysis of a Gaussian ARMA sequence, gives ML estimation and other diagnostic checks, can be used.

One can generate a parametric family by assuming that the d.f. H is a function of parameters $\theta, \theta \in \Theta$. Cauchy, Logistic and Weibull time series can be generated by taking H^{-1} as $\tan(\pi p)$, $-\ln(p^{-1}-1)$, $1-\exp(-\mu(x-\beta)^\gamma)$, $x \geq \beta$ respectively. The Weibull series includes the exponential model with $\gamma = 1$. Likelihood can be written in such a case in terms of likelihood of a Gaussian ARMA sequence.

Thus, models defined above have an operational convenience. Properties such as mixing are also easy to establish. However, it seems that they have not been fully explored in statistical analysis of non-Gaussian time series.

Remark 3.5.1 A number of models discussed above form state-space models or generalized linear models, cf. Fahrmeir and Tutz (2004).

Remark 3.5.2 Elliot et al. (1995) and Cappé et al. (2005) extensively discuss Hidden Markov Models on a general state-space along with their applications.

3.6 Goodness-of-Fit Procedures

So far, we have seen time series models which have a particular distribution of interest, as the unique stationary distribution. It is therefore important to have goodness of fit tests for the proposed stationary distribution. Such procedures can be applied to models in Chap. 2 also.

We discuss one such procedure which is based on the classes, which form a partition of the state-space S of the series. We can also consider the classes which form a partition of $S \times S$. This would allow us to test the bivariate distribution of two consecutive observations of the time series under consideration and in a sense test for the conditional distribution also, which actually defines the series. Broadly speaking, these tests include the well known Pearsonian chi-squared tests routinely applied in the case of i.i.d. observations.

Suppose we have T observations $X_1, X_2, \ldots, X_T$ from a stationary time series with P_θ as its probability measure when θ is the true parameter. Let Θ be the parameter space, an open subset of $\Re^p$. Let $Z(T, \theta) = (Z_1(T, \theta), Z_2(T, \theta), \ldots, Z_K(T, \theta))'$

be a $K \times 1$ random vector, where $Z_i(T, \theta)$ is a function of $X_1, X_2, \ldots, X_T$ and θ. We make the following assumptions.

A1. $\frac{1}{T}Z_i(T, \theta) \to 0$ a.s. for each i.

A2. For each i, the function $Z_i(T, \theta)$ is differentiable with respect to θ_j. Let $C(i, j) = \partial Z_i(T, \theta)/\partial \theta_j$. Then, for each i, j, $C_T(i, j)/T \to C(i, j)$, a.s. Let the matrix C be defined by $C = C(i, j)$.

A3. $\frac{1}{\sqrt{T}}Z(T, \theta) \overset{\mathscr{D}}{\to} N_K(0, \Sigma)$.

A4. Let, for each θ, $V(\theta)$ be a positive semi-definite matrix. For each (i, j), $V(i, j)$ is twice differentiable with respect to θ.

A5. For each θ, rank$(V(\theta)) = r > p$ and rank$(C'V_T^+C) = p$.

Let A^+ denote the Moore–Penrose g-inverse of a matrix A. Define

$$X^2(\theta) = \frac{1}{T}Z(T, \theta)'V(\theta)^+Z(T, \theta).$$

Let $G_{ij}(\theta) = \frac{\partial^2 X^2(\theta)}{\partial \theta_i \partial \theta_j}$. Let the matrix $G_T(\theta)$ be defined by $G_T(\theta) = ((G_{ij}(\theta)))$.

A6. There exists a neighborhood $N(\theta)$ of θ and a random variable $Y_t(\theta)$ such that, for all (i, j),

$$\sup_{\eta \in N(\theta)} |G_{ij}(\eta) - G_{ij}(\theta)| \le |\eta - \theta|Y_T(\theta)$$

and $Y_T(\theta)/T \to Y(\theta)$ a.s., where $0 \le E(Y(\theta)) < \infty$. Let

$$H_T(\theta, j) = \frac{\partial X^2(\theta)}{\partial \theta_j}, \quad j = 1, 2, \ldots, p.$$

Let $H_T(\theta) = \big(H_T(\theta, 1), H_T(\theta, 2), \ldots, H_T(\theta, p)\big)$. Under the above assumptions, Rajarshi (1987) shows the that

1. $\frac{1}{T}H_T(\theta, j) \to 0$, a.s.,
2. $\frac{G_T(\theta)}{T} \to 2C'V^+C$, a.s.

By making arguments similar to those made in the proof of Theorem 1.5.1, it can be shown that there exists a sequence of estimators $\hat{\theta}$ such that

$$P[H_T(\hat{\theta}) = 0] \to 1 \quad \text{and} \quad \hat{\theta} \to \theta \text{ a.s.}$$

Moreover,

$$\sqrt{T}(\hat{\theta} - 0) \overset{\mathscr{D}}{\to} N_p(0, DC'V^+\Sigma V^+CD), \quad \text{where} \quad D = (C'V^+C)^{-1}.$$

In particular, if $\Sigma = V(\theta)$, $X^2(\hat{\theta}) \overset{\mathscr{D}}{\to} \chi^2_{r-p}$.

A Pearson-type chi-squared test based on the frequencies of various classes is constructed as follows. Let $A_1, A_2, \ldots, A_K$ be a partition of S. It is assumed that

$P_\theta(X_1 \in A_i) > 0$, $i = 1, 2, \ldots, K$. Let $N_T(i) = \sum_{t=1}^{T} I[X_t \in A_i]$ be the number of observations that belong to the class A_i, $i = 1, 2, \ldots, K$. By taking $Z_i(T, \theta) = N_T(i) - P_\theta(X_1 \in A_i)$, $i = 1, 2, \ldots, K$ and assuming that the assumptions hold, we get a χ^2 statistic with $(K - 1 - p)$ d.f., cf. (2.2). In practice, it is more convenient to estimate the variance-covariance matrix of the observed frequencies following the procedure in Example 2.2.6.

Rajarshi (1987) discusses applications to a goodness of fit test of a first order Markov sequence, where the test is based on the first K conditional moments given the last observation.

Apart from the formal goodness-of-fit tests based on frequency distributions (joint or marginal), a number of other procedures need to be attempted for model validation. If the first order Markov property has been assumed, in most of the models that we have discussed, the sample ACF should have an exponential decay and the Sample PACF should have only a spike of order one. Let us assume that a parametric model admits a score function, say, $S_t(\theta) = \partial \ln f(X_t|X_{t-1})/\partial\theta$. Then, let $\tilde{S}_t = S_t(\hat{\theta})/\{E[S_t^2(\theta)|X_{t-1}]\}_{\hat{\theta}}$. The graph of S_t's against the time parameter t should resemble that of a white noise, if the model is a good approximation. If the score function is not defined or if it is complicated, one can plot $\{X_t - E_{\hat{\theta}}[X_t \mid X_{t-1}, \ldots, X_0]\}/\{Var_{\hat{\theta}}(X_t \mid X_{t-1}, \ldots, X_0)\}^{1/2}$. Similarly, ACF and PACF of $\tilde{S}_t$'s is likely to reveal departures from assumptions of the model.

References

Abraham, B., Balakrishna, N.: Inverse gaussian autoregressive models. J. Time Ser. Anal. **20**, 605–618 (1999)

Adke, S.R., Balakrishna, N.: Estimation of the mean in some stationary Markov sequences. Commun. Stat. Theor. Methods **21**, 137–160 (1992)

Al-Osh, M.A., Alzaid, A.A.: First order integer valued auto-regressive (INAR(1)) process. J. Time Ser. Anal. **8**, 261–275 (1987)

Al-Osh, M.A., Alzaid, A.A.: Integer valued moving average (INMA) process. Statistical Hefte **29**, 281–300 (1988)

Al-Osh, M.A., Alzaid, A.A.: Binomial auto-regressive moving average models. Commun. Stat. Stoch. Model **7**, 261–282 (1991)

Al-Osh, M.A., Alzaid, A.A.: First order auto-regressive time series with negative binomial and geometric marginals. Commun. Stat. Theor. Methods **21**, 2483–2492 (1992)

Alzaid, A.A., Al-Osh, M.A.: An integer-valued p-th order auto-regressive structure(INAR(p)) process. J. Appl. Probab. **27**, 314–324 (1990)

Arnold, B.C., Hallett, J.T.: A characterization of the pareto process among stationary stochastic processes of the form $X_n = min(X_{n-1}, Y_n)$. Stat. Probab. Lett. **8**, 377–380 (1989)

Athreya, K.B., Pantula, G.S.: Mixing properties of harris chains and auto-regressive processes. J. Appl. Probab. **23**, 880–892 (1986a)

Athreya, K.B., Pantula, G.S.: A note on strong mixing ARMA processes. Stat. Probab. Lett. **4**, 187–190 (1986b)

Balakrishna, N., Jacob, T.M.: Parameter estimation in minification processes. Commun. Stat. Theor. Methods. **32**, 2139–2152 (2003)

Balakrishna, N., Lawrence, A.J.: Development of product autoregressive models. J. Indian Stat. Assoc. **50**, 1–22 (2012)

Balakrishna, N., Nampoothiri, K.: A cauchy auto-regressive model and its applications. J. Indian Stat. Assoc. **41**, 143–156 (2003)

Balakrishna, N., Shiji, K.: A markovian weibull sequence generated by product autoregressive model and its statistical analysis. J. Indian Soc. Probab. Stat. **12**, 53–67 (2010)

Benjamin, M., Rigby, R., Stasinopoulos, M.: Generalized autoregressive moving average models. J. Am. Stat. Assoc. **98**, 214–223 (2003)

Berchtold, A., Raftery, A.E.: Mixture transition distribution model for high-order markov chains and non-gaussian time series. Stat. Sci. **17**, 328–356 (2002)

Block, H.W., Langberg, N.A., Stoffer, D.S.: Time series models for non-gaussian processes, In: Topics in Statistical Dependence, Lecture Notes-Monograph Series, vol. 16, pp. 69–83. Institute of Mathematical, Statistics, Hayward (1990).

Brockwell, P.J., Davis, R.A.: Time Series: Theory and Methods. Springer, New York (1987)

Cappé, O., Moulines, E., Ryden, T.: Inference in Hidden Markov Models. Springer, New York (2005)

Drost, F.C., van den Akker, R., Werker, B.J.M.: Efficient estimation of auto-regressive parameters and innovation distribution for semi-parametric integer-valued AR(p) processes. J. Roy. Stat. Soc.: Ser. B, 71, 467–485 (2009).

Du, J.D., Li, Y.: The integer-valued auto-regressive (INAR(p)) model. J. Time Ser. Anal. **12**, 129–142 (2001)

Elliot, R.J., Aggaoun, L., Moore, J.B.: Hidden Markov Models: Estimation and Control. Springer, New York (1995)

Fahrmeir, L., Tutz, G.: Multivariate statistical modeling based on generalized linear models. Springer, New York (2004)

Feigin, P.D., Resnick, S.I.: Estimation of autoregressive processes with positive innovation. Commun. Stat. Stoch. Models **8**, 479–498 (1992)

Gaver, D.P., Lewis, P.A.W.: First-order autoregressive gamma sequences and point processes. Adv. Appl. Probab. **12**, 727–745 (1980)

Jacobs, P.A., Lewis, P.A.W.: A mixed auto-regressive-moving-average exponential sequence and point process (EARMA(1,1)). Adv. Appl. Probab. **9**, 87–104 (1977)

Jacobs, P.A., Lewis, P. A.W.: Discrete time series generated by mixtures I: correlational and run properties. J. Roy. Stat. Soc.: Ser. B, 40, 94–105 (1978a).

Jacobs, P.A., Lewis, P.A.W.: Discrete time series generated by mixtures II: asymptotic properties. J. Roy. Statist. Soc.: Ser. B, 40, 222–228 (1978b).

Jacobs, P.A., Lewis, P.A.W.: Stationary discrete auto-regressive-moving average time series generated by mixtures. J. Time Ser. Anal. **4**, 19–36 (1983)

Lewis, P.A.W., McKenzie, E.D.: Minification processes and their transformations. J. Appl. Probab. **28**, 45–57 (1991)

MacDonald, I., Zucchini, W.: Hidden Markov and Other Models for Discrete Valued Time Series. Chapman and Hall, London (1997)

McCabe, B.P.M., Martin, G.M., Harris, D.: Efficient probabilistic forecasts for counts. J. Roy. Stat. Soc.: Ser. B 73, 239–252 (2011).

McKenzie, E.: Some simple models for discrete variate time series. Water Res. Bull. **21**, 645–650 (1985)

McKenzie, E.: ARMA processes with negative binomial and geometric marginals. Adv. Appl. Probab. **18**, 679–705 (1986)

McKenzie, E.: Some ARMA models for dependent sequences of poisson counts. Adv. Appl. Probab. **20**, 822–835 (1988)

McKenzie, E.: Discrete variate time series. In: Shanbhag, D.N., Rao, C.R. (eds.) Stochastic Processes: Modeling and Simulation Handbook of Statistics, vol. 21, pp. 573–606. North-Holland, Amsterdam (2003)

Pillai, R.N., Satheesh, S.: α-inverse gaussian distributions. sankhyā A. 54, 288–290 (1992).

Raftery, A.E., Tavare, S.: Estimating and modeling repeated patterns in high order markov chains with mixture transition distribution model. Appl. Stat. **43**, 179–199 (1994)

Rajarshi, M.B.: Chi-squared type goodness of fit tests for stochastic models through conditional least squared estimation. J. Ind. Statist. Assoc. **24**, 65–76 (1987)

Tavares, L.V.: An exponential markovian stationary processes. J. Appl. Probab. **17**, 1117–1120 (1980), Addendum. J. Appl. Probab. **18**, 957 (1980)

Weiß, C.H: A new class of autoregressive models for time series of binomial counts. Commun. Stat. Theor. Methods 38, 447–460 (2009).

Chapter 4
Estimating Functions

Abstract In this chapter, we discuss methods of estimation of parameters which assume that the conditional expectation and conditional variance of an observable given the past observations have been specified. These constitute semi-parametric methods for stochastic models. We begin with Conditional Least Squares estimation. Then, we discuss estimating functions in some details. The basic set of estimating functions can be conditionally uncorrelated or correlated. Optimality results for both these cases have been established. Asymptotic distribution of the estimator obtained from estimating equations is stated. Finally, we deal with methods of construction of confidence intervals based on estimating functions.

4.1 Conditional Least Square Estimation

The method of Conditional Least Squares (CLS) estimation (Klimko and Nelson 1978) is an extension of the Least Squares estimation in a regression model to stochastic processes. Let Y_t be a function of $(X_0, X_1, \cdots, X_t)$ such that $\mathrm{Var}(Y_t) < \infty$. Let $H_t = Y_t - E[Y_t|X_0, X_1, \cdots, X_{t-1}]$, $t = 1, 2, \cdots$. In CLS estimation, the estimator of θ is the one which minimizes $\sum_{t=1}^{T} H_t^2$. If θ is a $p \times 1$ vector of real parameters and if the conditional expectation $E[Y_t|X_0, X_1, \cdots, X_{t-1}]$ is differentiable in θ, the CLS Estimators can be obtained by solving

$$\sum_{t=1}^{T}(Y_t - E[Y_t|X_0, X_1, \cdots, X_{t-1}])\frac{\partial E[Y_t|X_0, X_1, \cdots, X_{t-1}]}{\partial \theta_i} = 0,$$

$$i = 1, 2, \cdots, p. \qquad (4.1)$$

We illustrate the CLS estimation methods by few examples.

M. B. Rajarshi, *Statistical Inference for Discrete Time Stochastic Processes*,
SpringerBriefs in Statistics, DOI: 10.1007/978-81-322-0763-4_4,
© The Author(s) 2012

Example 4.1.1 Poisson Markov Sequence (Example 2.2.7 continued)

We recall that $E[X_t|X_0, X_1, \cdots, X_{t-1}] = pX_{t-1} + \lambda$ and take $Y_t = X_t$. The CLS Estimation equations for (p, λ) are given by

$$\sum_{t=1}^{T}(X_t - pX_{t-1} - \lambda)X_{t-1} = 0$$

$$\sum_{t=1}^{T}(X_t - pX_{t-1} - \lambda) = 0,$$

which can be solved very easily.

Example 4.1.2 Standard linear Auto-regressive sequences.

The CLS Estimators of the auto-regressive parameters turn out to be the same as those obtained from the sample Yule-Walker equations.

The CAN property of CLS estimators, under regularity conditions, has been established in Klimko and Nelson (1978), also see Hall and Heyde (1980), Sect. 6.3. Both the proofs apply versions of martingale convergence theorem and martingale CLT.

The CLS method can be easily applied to models discussed in the Chap. 3, for which the conditional first moment is linear in parameters, in particular for INAR(1) models of Sect. 3.1 and models of Sect. 3.2. Weiß (2009) discusses CLS method of estimation for Binomial AR(p) models. Closed form for CLS estimators may not be available for some models and iteration procedures may be required. It is to be noted that in some models, the first conditional moment may not be a function of all the parameters of a interest.

The CLS estimation approach does not take into account the conditional variance of Y_t, and thus corresponds to the Ordinary LS (OLS) estimation. The approach based on estimating functions that we discuss next, corrects for this shortcoming of the CLS estimation. However, it requires a statistician to specify the conditional variance of Y_t given the past.

4.2 Optimal Estimating Functions

We begin with the observation that the following widely used methods of estimation of parameters involve solving simultaneous equations in unknown parameters. In the single parameter case, we have

(i) $\frac{\partial \log L}{\partial \theta} = 0$ (the likelihood equation),

(ii) $\sum_{t=1}^{T}(X_t - \beta z_t)z_t = 0$ (the least square equation for the regression parameter β, z_t is the regressor),

(iii) $\overline{X} - \mu(\theta) = 0$ (the method of moments estimator).

The CLS method described in the last section is an important example of an estimating equation , particularly useful in stochastic models.

Let X be an observable with its probability distribution P and let $\mathscr{P} = \{P\}$ denote the underlying family of distributions of the observable X.

Definition 4.2.1 The function H (a $p \times 1$ vector) of observations X and θ, a $p \times 1$ vector of parameters, such that the estimator of θ is obtained by solving the simultaneous equations $H = 0$, is called as an estimating function . The equation $H = 0$ is called as an estimating equation .

Definition 4.2.2 An estimating function H is said to be *unbiased* , if $E(H) = 0, \ \forall \ P \in \mathscr{P}$.

Definition 4.2.3 Let $p = 1$. An estimating function H is said to be *regular* , if for all $P \in \mathscr{P}$,it satisfies

  (i) $E(H) = 0$ and $E(H^2) < \infty$
  (ii) $0 < E \left| \frac{\partial H}{\partial \theta} \right| < \infty$ and $E \frac{\partial H}{\partial \theta}$ does not vanish.
  (iii) $\frac{\partial}{\partial \theta} \int H(x; \theta) f(x; \theta) dx = \int \frac{\partial}{\partial \theta} [H(x; \theta) f(x; \theta)] dx$.

Let $\mathscr{H} = \{H | H$ is a regular estimating function$\}$ be the class of all regular estimating functions. When is $H^* \subset \mathscr{H}$ optimal? We would like to have Var(H) to be small. Also, the estimating function H should be sensitive to changes in θ. Durbin (1960) and Godambe (1960) thus propose the following definition of an optimal estimating function .

Definition 4.2.4 (Durbin 1960; Godambe 1960). An estimating function $H^* \in \mathscr{H}$ is said to be optimal in $\mathscr{H}$, if H^* minimizes $\dfrac{V(H)}{\left[E \left(\frac{\partial H}{\partial \theta} \right) \right]^2}$,i.e.,

$$\frac{E(H^{*2})}{\left[E \left(\frac{\partial H^*}{\partial \theta} \right) \right]^2} \leq \frac{E(H^2)}{\left[E \left(\frac{\partial H}{\partial \theta} \right) \right]^2}, \quad \forall \ H \in \mathscr{H}.$$

Example 1. Let $X \sim N(\theta, 1)$ and let $H_1 = X - \theta$. We have $\partial H_1 / \partial \theta = -1$. Let $H_2 = (X - \theta)^2 - 1 \Rightarrow E(H_2) = 0$. But, $\frac{\partial H_2}{\partial \theta} = -2(X - \theta)$ which implies that $E \left(\frac{\partial H_2}{\partial \theta} \right) = 0$ and thus $H_2 \notin \mathscr{H}$.

Example 2. Let us consider a multinomial random vector (Y_1, Y_2, Y_3, Y_4) with $\sum\limits_{i=1}^{4} Y_i = n$ and the cell probabilities given by $\frac{1-\theta}{4}$, $\frac{1+\theta}{4}$, $\frac{1-\theta}{4}$, $\frac{1+\theta}{4}$. Then, $E \left[\frac{Y_1}{n} \right] = \frac{1-\theta}{4}; E \left[\frac{Y_2}{n} \right] = \frac{1+\theta}{4}$. Consider the estimating functions

$$H_1 = \frac{Y_1}{n} - \frac{1 - \theta}{4}; \quad H_2 = \frac{Y_2}{n} - \frac{1 + \theta}{4}.$$

Similarly, H_3 and H_4 can be obtained from Y_3 and Y_4 respectively. Then, $\left[E\left(\frac{\partial H_i}{\partial \theta}\right)\right]^2 = \frac{1}{16}$, $i = 1, \cdots, 4$. Thus, in this case, variances of the estimating functions allow us to choose between H_i, $i = 1, \cdots, 4$. The estimating function $H_1 + H_2 = \frac{Y_1 + Y_2}{n} - \frac{1}{2}$ is, however, free of θ and cannot be used for estimating θ. It is completely insensitive to changes in θ and it is not a regular estimating function.

Explanations regarding Godambe-Durbin criterion

In the scalar parameter case, assuming that the asymptotic works, $\dfrac{V(H)}{\left[E\left(\frac{\partial H}{\partial \theta}\right)\right]^2}$ is the variance of the asymptotic normal distribution of $\sqrt{T}(\hat{\theta}_T - \theta)$, where $\hat{\theta}_T$ is a consistent solution of $H = 0$, see Sect. 4.5 below. Thus, variance of asymptotic normal distribution of the estimator obtained by solving the optimal estimating equation $H = 0$ is smallest among variances of the asymptotic distribution of consistent solutions obtained by solving competing of the regular estimating equations. Godambe and Kale (1991) have pointed out convergence of the Newton-Raphson procedure for obtaining a solution of the optimal estimating equation is faster than that of the Newton-Raphson procedure for obtaining a solution of any competing regular estimating equation.

Remark 4.2.1 If H is a regular estimating function and if $H_1 = cH$, where c is a differentiable function of θ not involving any random variables, then

$$\frac{E(H^2)}{\left[E\left(\frac{\partial H}{\partial \theta}\right)\right]^2} = \frac{E(H_1^2)}{\left[E\left(\frac{\partial H_1}{\partial \theta}\right)\right]^2}.$$

Thus, the objective function does not change, if we multiply H by a constant and an optimal estimating function is unique up to a constant multiplier.

Semi-parametric models

In some situations, joint distributions of observables are not known, however, moments and covariances of certain elementary estimating functions can be known. Let H_i be a function of observations and θ, such that $E[H_i] = 0$ for all i and that each H_i is a regular estimating function. Further, $\text{Cov}(H_i, H_j) = 0$ for $i \neq j$, for all $P \in \mathscr{P}$. The estimating functions H_i's are known as uncorrelated elementary estimating functions and we seek to obtain a best linear combination $\sum_i w_i H_i$. Let $\mathscr{H}$ be the class of all estimating functions of linear combinations of $H_1, \cdots, H_n$, i.e.,

$$\mathscr{H} = \left\{ H = \sum_i w_i H_i \mid w_i\text{'s are differentiable functions in } \theta \right.$$

$$\left. \text{which do not involve any random variables} \right\}.$$

We seek an optimal estimating function in $\mathcal{H}$ as per the optimality criterion given in Definition 4.2.4. We observe that, in view of the assumption of uncorrelatedness and regularity assumptions,

$$\frac{E(H^2)}{\left[E\left(\frac{\partial H}{\partial\theta}\right)\right]^2} = \frac{\sum w_i^2 E(H_i^2)}{\left[\sum w_i E\left(\frac{\partial H_i}{\partial\theta}\right)\right]^2}.$$

Let $H_i' = \partial H_i/\partial\theta$. It is easy to see that the optimum weights w_i's which minimize the above, satisfy

$$w_i = \frac{\sum w_j^2 E(H_j^2)}{\sum w_j E(H_j')}\frac{E(H_i')}{E(H_i^2)}.$$

Since constant multipliers do not change a regular estimating function, we take the optimal weights as

$$w_i^* = \frac{E(H_i')}{E(H_i^2)}.$$

Thus, we conclude that $\sum H_i \dfrac{E(\partial H_i/\partial\theta)}{E(H_i^2)}$ is optimal in $\mathcal{H}$.

Example 4.2.1 Estimation of common mean

Let $\overline{Y}_1$, $\overline{Y}_2$ be uncorrelated sample means, based on n_1 and n_2 independent observations respectively. Suppose that θ is the common mean of the two populations. Let the population variances σ_1^2 and σ_2^2 be known. Now, for every a, $a\overline{Y}_1 + (1-a)\overline{Y}_2$ is unbiased for θ. The optimal a which minimizes the variance is given by $a = \left(\frac{n_1}{\sigma_1^2}\right)\Big/\left(\frac{n_1}{\sigma_1^2} + \frac{n_2}{\sigma_2^2}\right)$. The optimal estimating function with $H_1 = \overline{Y}_1 - \theta$, $H_2 = \overline{Y}_2 - \theta$ and $H = w_1 H_1 + w_2 H_2$ leads to the same estimator.

Example 4.2.2 Linear Regression

Suppose that $H_i = Y_i - \theta x_i$, where Y_i's are independent responses with means θx_i, $i = 1, 2, \cdots, n$. Let $V(H_i) = \sigma^2 V_i$. Here, $x_1, \cdots, x_n$ are constants (regressors) and V_i is known for each i. The optimal estimating equation is given by

$$\sum (Y_i - \theta x_i)\frac{x_i}{\sigma^2 V_i} = 0.$$

The particular cases (1) $V_i \equiv 1$, (2) $V_i = x_i$ and (3) $V_i = x_i^2$ respectively correspond to Optimal Ordinary Least Squares , Ratio estimator and the mean of ratios, which are BLUEs under the assumptions on V_i's in each case.

Example 4.2.3 Simultaneous auto-regressive models (Cliff and Ord 1981)

Whittle (1954) proposed the following model in connection with modeling data on uniformity trial on wheat. Let $\{\epsilon_t\}$ be a sequence of i.i.d. random variables with mean 0 and variance σ^2. Let

$$X_t = \theta \sum_j A_{tj} X_j + \epsilon_t,$$

where the sum over j corresponds to appropriate neighbours of t. Here, $A_{tt} = 0\ \forall t$. For the sake of simplicity, we assume that for all t, $A_{t,t+1} = A_{t,t-1} = 1$ and that the rest of A_{tj}'s are all zeroes. Here, if we take $H_t = X_t - \theta(X_{t+1} + X_{t-1})$, $E(\partial H_t/\partial\theta) = 0$ so that H_t is not a regular estimating function.

Naik-Nimbalkar (1996) suggests that we set

$$H_t = \big(X_t - \theta(X_{t+1} + X_{t-1})\big)\big(X_{t+1} - \theta(X_{t+2} + X_t)\big), \ t = 2, \cdots, T1.$$

We observe that H_t is a regular estimating function and that H_t's are uncorrelated (though they are not independent random variables). Under the assumptions that the sequence $\{X_t, t \geq 1\}$ is stationary, we have $E(\partial H_t/\partial\theta) = c\ \forall t$. Further, $\text{Var}(H_t) = \sigma^4$. Hence, from the above discussion, it follows that $\sum H_t$ is an optimal estimating function in the class of estimating functions given by $\sum w_t H_t$. Ord (1975) derives the same estimating function from the theory of weighted least squares . He further shows that the ordinary least square estimator is inconsistent. The above examples show that when the variances do not depend upon the parameter of interest, the theory of optimal estimating function gives the same solution as the weighted least squares estimators. In the next example, variances are functions of parameters in which case, the estimating function leads to more reasonable estimators than the weighted least square estimators.

Example 4.2.4 A non-linear regression model

Let $H_i = Y_i - e^{\theta x_i}$, $E(Y_i) = e^{\theta x_i}$, $\text{Var}(H_i) = e^{\theta x_i}$ for all $i = 1, 2, \cdots, T$. Models satisfying these conditions include the class of Poisson distributions of $Y_i \sim Poisson(e^{\theta x_i}.)$ In this case, $E\left(\frac{\partial H_i}{\partial\theta}\right) = x_i e^{\theta x_i}$. Thus, the optimal estimating function is given by

$$\sum (Y_i - e^{\theta x_i}) x_i = 0.$$

The above coincides with the score function, if Y_i has a Poisson distribution. The above equation is also obtained by methodology of the Generalized Linear Models. It needs to be pointed out that when the variances are functions of the unknown parameter θ, the weighted least square estimation involves minimizing

$$\sum \frac{(Y_i - e^{\theta x_i})^2}{e^{\theta x_i}}.$$

The corresponding estimating equation is given by

$$\sum (Y_i - e^{\theta x_i}) x_i + \sum \frac{(Y_i - e^{\theta x_i})^2 x_i}{e^{\theta x_i}},$$

which does not agree with the score function for the Poisson distribution. Further, in general, it is not an unbiased estimating function. Thus, the theory of estimating functions attempts to combine strengths of the likelihood approach and the least square approach.

We conclude our discussion by stating two important properties of an optimal estimating function in $\mathcal{H}$. Proofs of both the parts are easy.

Theorem 4.2.1 *Suppose that the regularity conditions hold.*

(i) An Estimating function H^* is optimal in $\mathcal{H}$, if and only if,

$$Cor\left(H, \frac{\partial \log L}{\partial \theta}\right) \le Cor\left(H^*, \frac{\partial \log L}{\partial \theta}\right), \quad \forall \ H \in \mathcal{H}.$$

(ii) An estimating function H^* is optimal in $\mathcal{H}$, if and only if,

$$E\left(H - \frac{\partial \log L}{\partial \theta}\right)^2 \ge E\left(H^* - \frac{\partial \log L}{\partial \theta}\right)^2, \quad \forall \ H \in \mathcal{H}.$$

Thus, the optimal estimating function is nearest to the score function, which is optimal in a larger class of regular estimating functions.

4.3 Estimating Functions for Stochastic Models

We now discuss estimating functions for semi-parametric stochastic models. Let X denote the observations. Let $\mathcal{F}_{t-1}$ be the σ-field generated by the random variables $(X_0, X_1, \cdots, X_{t-1})$. We note that $\mathcal{F}_{t-1} \subset \mathcal{F}_t$. Suppose that there exist $H_1, H_2, \cdots, H_T$, where each $H_t = H_t(X, \theta)$ is a function of observation X_t and a scalar parameter θ, such that

(i) $E(H_t | \mathcal{F}_{t-1}) = 0$, $Var(H_t) < \infty$.
(ii) $E\left[\left|\frac{\partial H_t}{\partial \theta}\right|\right] < \infty$ and $E \frac{\partial H_t}{\partial \theta}$ does not vanish.
(iii) Each H_t satisfies the following regularity condition

$$\frac{\partial}{\partial \theta} \int (H_t(x, \theta) | \mathcal{F}_{t-1}) \ f_\theta(x | \mathcal{F}_{t-1}) dx = \int \frac{\partial [H_t(x, \theta) | \mathcal{F}_{t-1}) f_\theta(x | \mathcal{F}_{t-1})]}{\partial \theta} dx,$$

with an obvious notation. Here and in the sequel, the integral should be interpreted as a sum when the underlying random variables are discrete.

(iv) $V_t = \text{Var}(H_t \mid \mathscr{F}_{t-1})$ is positive $\forall\, t$ and its functional form is known (it may involve θ).

Elementary estimating functions satisfying the above conditions are known as *orthogonal* regular estimating functions . Consider the class of estimating functions

$$\mathscr{H} = \left\{ H = \sum_{t=1}^{T} W_t H_t \mid W_t \text{ is a function of } (X_0, X_1, \cdots, X_{t-1}) \text{ and } \theta \right\}.$$

It is also assumed that W_t is a differentiable function of θ. As before, an optimal estimating function in $\mathscr{H}$ is the one that minimizes $E(H^2)/\left[E\left(\frac{\partial H}{\partial \theta}\right) \right]^2$.

Theorem 4.3.1 (Godambe 1985) *Suppose $\{H_t, t = 1, \cdots, T\}$ is a collection of orthogonal and regular elementary estimating functions. Then, the optimal estimating function is given by $H^* = \sum_{t=1}^{T} W_t^* H_t$, where*

$$W_t^* = \frac{E\left(\left.\frac{\partial H_t}{\partial \theta}\right| \mathscr{F}_{t-1}\right)}{\text{Var}(H_t|\mathscr{F}_{t-1})}, \quad t = 1, \cdots, T.$$

Proof We have

$$E\left[\frac{\partial H}{\partial \theta}\right] = E\left[\sum_t E\left(W_t \frac{\partial H_t}{\partial \theta} \middle| \mathscr{F}_{t-1}\right)\right]$$

$$= E\left[\sum_t W_t W_t^* H_t^2\right] = \text{Cov}\left(\sum_t W_t H_t, \sum_t W_t^* H_t\right),$$

in view of the orthogonality of H_t's and the fact that both W_t and W_t^* are measurable with respect to $\mathscr{F}_{t-1}$. Therefore, by the Cauchy-Schwarz inequality, we have

$$\left[E\left(\frac{\partial H}{\partial \theta}\right)\right]^2 \leq V(H)V(H^*).$$

But, it can be easily seen that $V(H^*) = E[\partial H^*/\partial \theta]$. Hence, $V(H)/[E\left(\frac{\partial H}{\partial \theta}\right)]^2 \geq V(H^*)/[E\left(\frac{\partial H^*}{\partial \theta}\right)]^2$.

Example 4.3.1 The Linear Auto-Regression Model

Let $X_t = \theta X_{t-1} + \epsilon_t$, where $\{\epsilon_t\}$ are independent random variables with mean 0 and variance σ^2. We note that ϵ_t is independent of $X_0, \cdots, X_{t-1}$. We then have $E(X_t|\mathscr{F}_{t-1}) = \theta X_{t-1}$ and $V(X_t|\mathscr{F}_{t-1}) = \sigma^2$. Now, let $H_t = X_t - \theta X_{t-1}$, $t = 1, 2, \cdots, T$. Then, it is easily seen that the estimator obtained by solving the optimal estimating function and the CLS Estimator coincide. This estimator is also the ML estimator, if the errors are independent $N(0, \sigma^2)$ random variables and if we agree to ignore the information in the initial observation.

We note that if the conditional variance of H_t is a function of constants and parameters (free from t also) only, the CLS equation coincides with the optimal estimating equation. Further, if V_t is a non-constant function of past observations, the optimal estimating function is better than the estimating function corresponding to the CLS estimation equation. This follows since the estimating function corresponding to CLS belongs to the class $\mathcal{H}$.

Example 4.3.2 Random Coefficient Auto Regressive (RCAR) model

Suppose that $\{\theta_t\}$ is a sequence of i.i.d. r.v.s such that $E(\theta_t) = \theta$ and $\mathrm{Var}(\theta_t) = \sigma_\theta^2$. Let $\{\epsilon_t\}$ be an i.i.d. sequence with $E(\epsilon_t) = 0$, $V(\epsilon_t) = \sigma^2$. We further assume that the random sequence $\{\epsilon_t\}$ is independent of the random sequence $\{\theta_t\}$.
The RCAR model is given by $X_t = \theta_t X_{t-1} + \epsilon_t$. Now, in view of the assumptions, we have $E(X_t|X_0, \cdots, X_{t-1}) = E(\theta_t X_{t-1}|X_0, \cdots, X_{t-1}) + E(\epsilon_t|X_0, \cdots, X_{t-1}) = X_{t-1}E(\theta_t|X_0, \cdots, X_{t-1}) = \theta X_{t-1}$.
Thus, we take

$$H_t = X_t - \theta X_{t-1}, \quad t = 1, 2, \cdots, T,$$

as elementary estimating functions. Then, $H_t^2 = (X_t - \theta X_{t-1})^2 = (X_t - \theta_t X_{t-1})^2 + [(\theta_t - \theta)X_{t-1}]^2 + 2(X_t - \theta_t X_{t-1})(\theta_t - \theta)X_{t-1}$. Therefore, $E(H_t^2|\mathscr{F}_{t-1}) = \sigma^2 + \sigma_\theta^2 X_{t-1}^2$. The optimal estimating function for θ is thus given by

$$H^* = \sum \frac{(X_t - \theta X_{t-1})X_{t-1}}{\sigma^2 + \sigma_\theta^2 X_{t-1}^2}.$$

The corresponding estimator is given by

$$\hat{\theta} = \sum_{t=1}^{T} \frac{X_t X_{t-1}}{\sigma^2 + \sigma_\theta^2 X_{t-1}^2} \bigg/ \sum_{t=1}^{T} \frac{X_{t-1}^2}{\sigma^2 + \sigma_\theta^2 X_{t-1}^2}. \tag{4.2}$$

The above discussion assumes that the two variances are known. If the "environmental" variance σ_θ^2 is too large as compared to σ^2, the above leads to the estimator $(1/T)\sum_t (X_t/X_{t-1})$.

If these two variances are unknown, we can proceed as follows. We note that $E\left[(X_t - \theta X_{t-1})^2|\mathscr{F}_{t-1}\right] = \sigma^2 + \sigma_\theta^2 X_{t-1}^2$, which, for a known θ, can be taken as a linear regression model with the two unknown variances as the intercept and the slope parameters and X_{t-1}^2's as regressors. This leads to the following two-stage estimator of θ.

1. Obtain the CLSE $\hat{\theta}(CLSE) = \frac{1}{T}\sum_{t=1}^{T} \frac{X_t X_{t-1}}{X_{t-1}^2}$.

2. Obtain the LSE of σ^2 and σ_θ^2 by regressing $(X_t - \hat{\theta}(CLSE)X_{t-1})^2$ on $\sigma^2 + \sigma_\theta^2 X_{t-1}^2$, $t = 1, 2, \cdots, T$.

Table 4.1 RMSE for three estimators in RCAR model

Sample sizes		50			500			1000		
θ	σ_θ^2	CLSE	EF	IT-EF	CLSE	EF	IT-EF	CLSE	EF	IT-EF
0.1	0.16	0.1537	0.1533	0.1525	0.0486	0.0477	0.0467	0.0355	0.0343	0.0340
−0.3	0.25	0.6815	0.2791	0.2408	0.0397	0.0375	0.0369	0.0403	0.0363	0.0358
0.8	0.30	0.1698	0.1626	0.1505	0.0665	0.0507	0.0466	0.0536	0.0346	0.0302
0.3	0.50	0.1929	0.1816	0.1833	0.0798	0.0560	0.0556	0.0627	0.0400	0.0400

3. If either of these is negative or 0, we take the CLSE as the final estimator of θ. If both are positive, replace the unknown variances by these two in (4.2). We call this as the EF estimator.

We have two options. We can either stop at the first iteration or we can continue with an iteration procedure. In the iteration procedure, successive iterations are carried out by taking Least Squares type estimators of the two variances obtained at the previous iteration, to compute the optimal estimator in this iteration.

The iteration procedure is terminated by using the usual stopping rules. This estimator is denoted by IT-EF estimator. Abdullah et al. (2011) compare these three estimators. It is assumed that $\epsilon_t \sim N(0, 1)$ and $\theta_t \sim N(\theta, \sigma_\theta^2)$. They consider the sample sizes 50, 500 and 1000 and carry out 1000 simulations for each combination. The Table 4.1, which is based on the Table 4.1 of Abdullah et al. (2011), gives the Root Mean Squared Errors (RMSEs) of the three estimators. It is concluded based on these simulation studies that both the one step EF estimator and the corresponding iterative procedure estimator perform better than the CLSE in each case and they perform considerably better, particularly for larger samples and for positive values of θ. The one step procedure seems to be working quite satisfactorily as compared to the iterative procedure. Abdullah et al. (2011) also compare the estimators of the two variances. Their simulations indicate that there is no significant gain in the RMSE of estimators of variances.

Adke and Balakrishna (1992) consider a special case of RCAR model wherein a random auto-regression coefficient has a discrete distribution on 0 and a constant $\beta \in (0, 1)$, so that $\theta < 1$. Further, the stationary distribution of the sequence is exponential with mean μ. This is the exponential AR(1) model due to Lawrence and Lewis (1981). This is a minification sequence model and as discussed in Sect. 3.3, we may assume that the parameter θ is known. Adke and Balakrishna (1992) show that the BLUE of μ has a smaller variance than the CLSE, though variances of their asymptotic normal distributions coincide.

4.4 Estimating Functions for a Vector Parameter

The following theorem proves optimality of the likelihood equation or the score function for a family of parametric models with p parameters. Let $\theta = (\theta_1, \theta_2, \cdots, \theta_p)'$ be a $p \times 1$ parameter, $\theta \in \Theta$, a p-dimensional open set in $\Re^p$.

Definition 4.4.1 An estimating function $H = (H_1, H_2, \cdots, H_p)'$, a function of observations and the parameters, is said to be *regular*, if for all $P \in \mathscr{P}$,

(i) $E(H_i) = 0$, $i = 1, 2, \cdots, p$.

(ii) $E(H_i^2) < \infty$ for all i.

(iii) $0 < E \left| \frac{\partial H_i}{\partial \theta_j} \right| < \infty$ for all i, j.

(iv) For all i, j,

$$\frac{\partial}{\partial \theta_j} \int H_i(x; \theta) f(x; \theta) dx = \int \frac{\partial}{\partial \theta_j} [H_i(x; \theta) f(x; \theta)] dx.$$

(v) The matrix $D(H) = ((E[\partial H_i / \partial \theta_j]))$ is non-singular.

We assume that the vector of score functions $S(\theta) = (\partial \ln L(\theta)/\partial \theta_i)$, $i = 1, 2, \cdots, p$ exists and also a regular estimating function with $I(\theta)$, the Fisher Information matrix, as its variance-covariance matrix. Let $V(H)$ denote the variance-covariance matrix of the random vector H.

Theorem 4.4.1 (Kale 1962). *For any regular estimating function H, the matrix $V(H) - D(H)'[I(\theta)]^{-1} D(H)$ is positive semi-definite.*

Proof In view of the regularity conditions,

$$D(H)_{ij} = E[\partial H_i / \partial \theta_j] = -E[H_i S_j] = -\text{Cov}(H_i, S_j).$$

Therefore, the matrix

$$\begin{bmatrix} V(H) & -D(H) \\ -D(H) & I(\theta) \end{bmatrix}$$

is positive semi-definite since it is the variance-covariance matrix of the $2p \times 1$ random vector $(H', S(\theta)')'$. It follows from the standard multivariate arguments that the matrix $V(H) - D(H)'[I(\theta)]^{-1} D(H)$ is positive semi-definite. In fact, it is the variance-covariance matrix of the random vector $H - D(H)' I(\theta)^{-1} S(\theta)$.

Definition 4.4.2 Optimal estimating function for a vector parameter

An estimating function H in the class of regular estimating functions is said to be optimal, if $V(H) = D(H)'[I(\theta)]^{-1} D(H)$.

Corollary *The Score function is optimal in the class of regular estimating functions.*

The Corollary follows since, under the regularity conditions, we have $V(H) = -D(H) = I(\theta)$. This leads to the optimality of the Score function. This property of the score function was proved by Godambe (1960) for a scalar parameter. Both the term estimating function and its theory have been in use for a long time, see McLeish and Small (1988), page 10, wherein it is also pointed out that the finite sample optimality of the score function was known to Barnard.

In stochastic models, we may not know the likelihood and our information may be limited to conditional moments of an observation given the past. Thus, we seek an optimal estimating function in an appropriate sub-class of regular estimating functions $\mathscr{H}$. Optimality of an estimating function is as defined below.

Definition 4.4.3 An estimating function $H^* \in \mathscr{H}$ is said to optimal in $\mathscr{H}$, if

$$D(H)V(H)^{-1}D(H)' - D(H^*)V(H^*)^{-1}D(H^*)'$$

is positive semidefinite for all $H \in \mathscr{H}$ and for every member of the family of distributions of X.

The above criterion for optimal estimating function has been given by Durbin (1960), Kale (1962), Bhapkar (1972), Godambe and Thompson (1989). We now give a generalization of Godambe's result (Theorem 4.3.1) to the case of many parameters. When the score function exists, the criterion is equivalent to the following.

Definition 4.4.4 An EF H^* is optimal in $\mathscr{H}$ if and only if

$$E[(S(\theta) - H^*)D(H)'] = E[D(H)(S(\theta) - H^*)'] = 0.$$

It can be also shown that, when the score function exists and the regularity conditions are satisfied, H^* is nearest to the score function $S(\theta)$ in the sense that $E[S(\theta - H)((S(\theta - H))'] - E[(S(\theta - H^*)((S(\theta - H^*))'] $ is a positive semidefinite or definite matrix. Godambe and Heyde (1987) further show that the confidence set based on a properly Studentised optimal estimating function has smallest volume in the class of confidence sets based on competing Studentised estimating functions, assuming that a CLT holds for all the estimating functions in the class $\mathscr{H}$.

In general, the score function may not exist and in such a case, optimality of an estimating function needs to be proved directly from the definition of optimality in the multi-parameter case. We proceed to prove such a result when we have a martingale structure of the following type.

Definition 4.4.5 Let $H_t(j)$, $t = 1, 2, \cdots, T$, $j = 1, 2, \cdots, K$ be a set of regular estimating functions such that

1. $E[H_t(j)|\mathscr{F}_{t-1}] = 0$.

2. $E[H_t(j)H_t(i)|\mathscr{F}_{t-1}] = 0$, $i \neq j$.

Then, the estimating functions $H_t(j)$'s are said to be *mutually orthogonal*.

A class of estimating functions $\mathscr{H} = \{H\}$ is defined as follows. Let $H = (H(1), H(2), \cdots, H(p))'$ be a vector estimating function such that

$$H(j) = \sum_{t=1}^{T} \sum_{i=1}^{K} W_t(i, j) H_t(i), \quad j = 1, 2, \cdots, p,$$

where $W_t(i, j)$ are random variables measurable with respect to $\mathscr{F}_{t-1}$.

Theorem 4.4.2 (Godambe and Thompson 1989) *An optimal estimating function H^* in $\mathscr{H}$ is given by*

$$W_t^*(i, j) = \frac{E\left[\left. \frac{\partial H_t(j)}{\partial \theta_i} \right| \mathscr{F}_{t-1} \right]}{E[H_t(j)^2 | \mathscr{F}_{t-1}]}.$$

Proof Arguing as before, it can be shown that

$$[E(\partial H(j)/\partial \theta_\ell)]^2 \leq \text{Var}(H(j)) \text{Var}(H_\ell^*).$$

Thus, it follows that the matrix

$$\begin{bmatrix} V(H) & D(H) \\ D(H)' & V(H^*) \end{bmatrix}$$

is non-negative definite. One can then show that, for the optimal estimating function H^*, we have $D(H^*) = H^*$, see Lemma 1 of Hwang and Basawa (2011).

The condition of mutual orthogonality appears to be somewhat restrictive, however, it is not. Let $\Sigma_t(\theta)$ be the conditional variance-covariance matrix of $(H_t(1), H_t(2), \cdots, H_t(p))'$. Assuming that $\Sigma_t(\theta)$ is a.s. a positive-definite matrix, we see that the p estimating functions obtained from $[\Sigma_t(\theta)]^{-1/2}(H_t(1), H_t(2), \cdots, H_t(p))'$ are mutually orthogonal. Frequently, we have $K = 1$.

Example 4.4.1 AR(2) model

Consider the AR(2) model given by $X_t = \theta_1 X_{t-1} + \theta_2 X_{t-2} + \epsilon_t$ with the usual set of assumptions. Then, ignoring σ^2 in the denominator in both the equations, we have the following estimating equations jointly optimal for θ_1, θ_2

$$\sum (X_t - \theta_1 X_{t-1} - \theta_2 X_{t-2}) X_{t-1} = 0$$

$$\sum (X_t - \theta_1 X_{t-1} - \theta_2 X_{t-2}) X_{t-2} = 0.$$

The above are the same as the sample Yule-Walker equations and the CLS equations.

Example 4.4.2 Poisson Markov sequence (Example 4.1.1 continued)

In this case, the optimal estimating equations are given by

$$\sum_t (X_t - pX_{t-1} - \lambda)\frac{X_{t-1}}{p(1-p)X_{t-1}+\lambda} = 0.$$

$$\sum_t (X_t - pX_{t-1} - \lambda)\frac{1}{p(1-p)X_{t-1}+\lambda} = 0.$$

The above equations need to be solved iteratively. The starting values can be taken as the CLS estimators.

Example 4.4.3 Chaotic Processes

Lele (1994) discusses applications of estimating function methodology to chaotic systems with measurement errors. The measurement error can be additive or multiplicative. Consider the Logistic map defined by $z_{t+1} = \theta z_t(1 - z_t)$. The observable process $\{X_t\}$ is defined by $X_t = z_t + \epsilon_t$, where $\{\epsilon_t, t \geq 1\}$ is a sequence of i.i.d. $N(0, \sigma^2)$ random variables. It follows that the elementary estimating function $X_{t+1} - \theta X_t(1 - X_t)$ is not an unbiased estimating function and that the correction $X_{t+1} - \theta X_t(1 - X_t) - \sigma^2$ yields an unbiased estimating function. Lele (1994) shows that, for known σ^2, the corresponding estimator is CAN. He also discusses the exponential map $z_{t+1} = z_t exp(\theta(1 - z_t))$. The corresponding multiplicative measurement error model is given by $X_t = V_t z_t$, where V_t is a sequence of i.i.d. lognormal random variables.

Example 4.4.4 Grouped data from finite Markov chains

Consider the grouped data from Markov chains (Sect. 2.5). Let $S(P)$ be the score function when we have observed all the individual one-step transitions for all the epochs under consideration. We may recall that the score function is a linear function of $N(i, j, t)$'s, where $N(i, j, t)$ is the number of one-step transitions from i to j at the epoch t. Let Y denote the grouped data as described in Sect. 2.5. The conditional expectation $E(S(P) \mid Y)$ is the optimal estimating function. However, it amounts to computing $E(N(i, j, t)|Y)$, which is too complicated to handle. McLeish (1984) considers the sub-class $\mathscr{H}_1$ of unbiased estimating functions which are linear functions of the observed proportions at various values of t. He projects the score function to the sub-class $\mathscr{H}_1$. It is further shown that there exist CAN solutions of such estimating equations. For a two-state Markov chain, it is possible to compute $E(N(i, j, t)|Y)$ explicitly, see Sect. 5 of McLeish (1984).

Remark 4.4.1 A standardized form of an estimating function H is defined as $H_s = [D(H)]^{-1}H$. In the multi-parameter case, there are a number of criteria of an optimal estimating function H_s^*. In terms of standardized form, the Matrix, Trace and Determinant optimality criteria, are respectively defined by

$$V(H_s) - V(H_s^*) \text{ is nnd, } \text{Trace}V(H_s) \geq \text{Trace}V(H_s^*) \text{ and } |V(H_s)| \geq |V(H_s^*)|,$$

where a matrix A is said to be nnd (non-negative definite), if it is either positive semi-definite or positive definite. Chandrasekhar and Kale (1984) have shown that these three criteria are equivalent in the sense that, if an estimating function is optimal with respect to one of the three, it is also optimal with respect to the remaining two. In a parametric set-up, the score function is thus optimal with respect to all the above criteria.

Our discussion here assumes regularity assumptions regarding a class of estimating functions as well the score function. For a discussion of estimating functions without such assumptions, we refer to McLeish and Small (1988) (who call such functions as "Statistical Inference Functions"), Godambe (1991) and Basawa et al. (1997). These books and monographs contain detailed accounts of theory and applications of estimating functions.

4.5 Confidence Intervals Based on Estimating Functions

In this section, we discuss confidence intervals for the unknown parameter, based on a CAN solution of an estimating equation and estimating function itself. The asymptotic normal distribution of a consistent solution of an estimating function has been proved by a large number of authors. More recent work is due to Chatterjee and Bose (2005) (the single parameter case) and Basawa and Hwang (2011) (the multiparameter case) under various sets of regularity assumptions. We write W_t and H_t for $W_t(\theta)$ and $H_t(\theta)$ respectively. Then,

$$\sqrt{T}(\hat{\theta}_T - \theta) \xrightarrow{\mathscr{L}} N\left(0, \frac{E[(W_1 H_1)^2]}{\left[E\left(\frac{\partial W_1 H_1}{\partial \theta}\right)\right]^2}\right)$$

where $\hat{\theta}_T$ is a consistent solution of $H = 0$.

Estimation of variance of the limiting distribution of $\hat{\theta}$.

Let $\hat{\theta}$ be a consistent solution of an estimating equation $H = 0$. Estimation of variance of $\hat{\theta}$ is itself of interest. It is further required to carry out tests of hypotheses or to construct confidence intervals for θ. The variance of the asymptotic distribution, in the stationary case, can be estimated as follows. We now explicitly write H_t and W_t as $H_t(\theta)$ and $W_t(\theta)$ respectively. Then, estimators of the required functions are given by

$$\hat{E}[(W_1 H_1)^2] = \frac{1}{T} \sum_{t=1}^{T} (W_t H_t)^2 (\hat{\theta}).$$

$$\hat{E}\left[\frac{\partial W_1 H_1}{\partial \theta}\right] = \sum_{t=1}^{T} \frac{\partial W_t H_t}{\partial \theta}\bigg|_{\theta = \hat{\theta}}.$$

Consistency of the above two estimators can be proved by assuming that (i) the function $\frac{\partial W_1 H_1}{\partial \theta}$ is continuously differentiable (ii) the function in a neighbourhood of the true parameter $\left| \frac{\partial^2 W_1 H_1}{\partial \theta^2} \right|$ is bounded by a function, the expectation of which is finite. The ergodic theorem completes the details. The standard approach of building a confidence interval for θ is based on the asymptotic normal distribution of $\hat{\theta}$ and a consistent estimator of the variance of the asymptotic normal distribution of $\sqrt{T}(\hat{\theta} - \theta)$. We call such a confidence interval (c.i.) as *Estimator based* c.i. . Let $Z_{\alpha/2}$ be such that

$$P[N(0, 1) > Z_{\alpha/2}] = \alpha/2. \tag{4.3}$$

The confidence interval based on an estimating function $\sum W_t H_t$ is constructed from the asymptotic pivotal given by

$$\frac{\sum W_t H_t}{\sqrt{\sum W_t^2 E(H_t^2 | \mathscr{F}_{t-1})}}.$$

We then equate the above pivotal to $Z_{\alpha/2}$ and solve the corresponding equations to get upper and lower limit of the required confidence interval. We call such a confidence interval as *EF based* c.i.

It has two important properties. It bypasses the estimation of θ. Secondly, the variance of an estimating function is estimated assuming that θ is known. This procedure and its variations have been discussed in Godambe (1985) among others, for stochastic models. The denominator can also be taken to be $\sqrt{\sum W_t^2 H_t^2}$.

We illustrate the two methods of confidence interval with the example of the auto-regressive parameter in an AR(1) model.

Example 4.5.1 Confidence interval for AR parameter (Example 4.3.1 continued)

To derive the Estimator based c.i., we recall from (1.2) that $\sqrt{T}(\hat{\theta} - \theta) \xrightarrow{\mathscr{L}} N\left(0, \frac{1}{1-\theta^2}\right)$. Let $\hat{\theta} = \sum X_t X_{t-1} / \sum X_{t-1}^2$ be the usual CLS estimator of θ. The standard Estimator based c.i. for θ is given by

$$\left(\hat{\theta} - Z_{\alpha/2} \frac{1}{\sqrt{T}} \frac{1}{\sqrt{1-\hat{\theta}^2}}, \ \hat{\theta} + Z_{\alpha/2} \frac{1}{\sqrt{T}} \frac{1}{\sqrt{1-\hat{\theta}^2}} \right).$$

In the AR(1) case, the optimal estimating equation is given by $\sum (X_t - \theta X_{t-1}) X_{t-1} = 0$. We take the EF pivotal as

$$G(\theta) = \frac{\sum (X_t - \theta X_{t-1}) X_{t-1}}{\sqrt{\sum X_{t-1}^2 (X_t - \theta X_{t-1})^2}}.$$

The EF based c.i. is conveniently obtained by solving the quadratic $G(\theta)^2 = Z_{\alpha/2}^2$. We simulated 5,000 samples from stationary AR(1) series with $T = 25$, $\theta = 0.8$ and $\sigma^2 = 1$. In the first case, the error distribution is $N(0, 1)$ and in the second case it is that of $Y - 1$, where Y is an exponential r.v. with mean 1. The nominal confidence coefficient is 95 %. The percentages of times the true value belongs to the confidence interval are 91.58 (Normal) and 90.54 (exponential) for the *Estimator based* c.i. and 95.36 (Normal) and 93.44 (exponential) for the *EF based* c.i.

In general, such upper and lower limits are obtained by solving more complicated equations by numerical methods such as Newton-Raphson.

Why does one anticipate that *EF based* c.i. performs better than the one based on the corresponding *Estimator based* c.i.? The EF pivotal has mean zero, unlike the random variable $\sqrt{T}(\hat{\theta} - \theta)$ except when $E(\hat{\theta}) = \theta$. Moreover, the asymptotic normality of the estimator is derived from the normality of the EF based pivotal. Typically, the estimator equals the suitably scaled EF plus a random term which converges to 0 in probability and a CLT is applied to the EF based pivotal. In this sense, as is often described, the EF pivotal is "more normal" than the estimator, see McLeish (1984), page 266.

4.6 Combining Correlated Estimating Functions

So far, we have assumed that the elementary estimating functions H_t's are orthogonal. However, in some situations, such an assumption may not be met. Suppose H_t, $t = 1, 2, \cdots, T$ is a collection of regular elementary estimating functions. Let $H = (H_1, H_2, \cdots, H_T)'$ be the random vector of these elementary functions. Let V denote the variance-covariance matrix of the random vector H and let

$$\mathbf{D} = [E(\partial H_1/\partial\theta), E(\partial H_2/\partial\theta), \cdots, E(\partial H_T/\partial\theta)]'.$$

Consider the class $\mathscr{H}$ of linear estimating functions $\sum_t w_t H_t$ where w_t's are constants which may possibly depend upon unknown parameters. Let w denote the $T \times 1$ vector $(w_1, w_2, \cdots, w_t)'$. The optimality criterion is the same as given in Sect. 2. Thus, an estimating function is optimal in the class $\mathscr{H}$, if it minimizes

$$\frac{w'Vw}{E\left[\sum \frac{\partial(w_t H_t)}{\partial\theta}\right]^2}$$

which reduces to

$$\frac{w'Vw}{(w'D)^2}.$$

From Rao (1965), page 48, it follows that $w'Va/(w'D)^2$ is minimum, when $w^* = V^{-1}D$. Then, the optimal estimating function in this class is given by

$$H^* = D'V^{-1}H \quad \text{with} \quad \text{Var}(H^*) = D'V^{-1}D.$$

If the sequence $\{H_t, t \geq 1\}$ is weakly stationary, it follows that $D = [E(\partial H_1/\partial\theta)]$ $E_{T\times 1}$, where $E_{a\times b}$ is an $a \times b$ matrix with unity everywhere. In general, both the matrices D and V may depend upon other unknown parameters and in such a situation, the scope of above optimal estimating function is limited.

We now discuss a special case, wherein the optimal estimating function simplifies considerably. If D is an eigen-vector of the matrix V with respect to a root λ, D is an eigen-vector of the matrix V^{-1} with respect to a root $1/\lambda$. Then, the optimal linear combination is given by $a^* = V^{-1}D = (1/\lambda)D$. We recall that optimality of an estimating function is not affected by multiplication by a constant. If further $D = E(\partial H_1/\partial\theta)E_{T\times 1}$, it follows that the simple estimating function $\sum_t H_t$ is optimal in $\mathcal{H}$.

Example 4.6.1 Exchangeable random variables

Suppose that $\{X_1, X_2, \cdots, X_T\}$ is a collection of exchangeable random variables with $E(X_t) = \theta$ and V as the variance-covariance matrix. Let $\rho = \text{Cor}(X_t, X_s)$, which is free from s, t in view of exchangeability. It follows that the row sums of V are constant and thus $E_{T\times 1}$ is an eigen-vector of the matrix V. Thus, $\bar{X}_t - \mu$ is an optimal estimating function for μ in the linear case. Of course, this is a restatement of the BLUE property of the sample mean in the equi-correlated case. In general, if $\{H_t, t \geq 1\}$ is an exchangeable process, it follows that $(1/T)\sum_t H_t$ is an optimal estimating function.

If the row sums of V^{-1} are constant and the process $\{H_t, t \geq 1\}$ is weakly stationary, it follows that the simple estimating function $\sum H_t$ is optimal. In the stationary AR(1) model with $E(X_1) = \mu$, the row sums of the inverse of variance-covariance matrix of $X_t - \mu$, $t = 1, 2, \cdots, T$ are almost the same, the first and the last row sums are different than the other row sums, which are the same. Thus, the estimating function $\sum_t (X_t - \mu)$ is nearly optimal and thus the sample mean may be regarded as a reasonable estimator. There are a number of processes which have this property, for example the Poisson Markov sequence has this property. A number of sequences discussed in Chap. 3 have the same ACF as that of the linear AR(1) model and the sample mean has such an (approximate) optimality property. Adke and Balakrishna (1992) verify that, in the case of the NEAR(1) model, the sample mean has the asymptotic variance same as that of the BLUE, which can be obtained from the theory of optimal estimating functions. In general, an optimal estimating function needs information on the correlation structure of the observed process and a modeler may not be willing to assume any structure. We may prefer a simple estimating function such as $\sum H_t$ to an optimal estimating function, which is difficult to obtain and to work with.

Example 4.6.2 Confidence interval for the mean of a stationary process.

Let $\{X_t, t \geq 1\}$ be a strictly and weakly stationary process. We first describe the *Estimator based* confidence interval. Under appropriate conditions on moments and strong mixing coefficients (Theorem 1.3.4), we have

$$\sqrt{T}(\overline{X} - \mu) \overset{\mathcal{D}}{\to} N(0, \sigma^2),$$

where

$$\sigma^2 = \text{Var}(X_1) + 2 \sum_{t=1}^{\infty} \text{Cov}(X_1, X_{t+1}).$$

A consistent estimator of σ^2 is given by

$$\hat{\sigma}_T^2 = \frac{1}{T} \sum_t (X_t - \overline{X})^2 + 2 \sum_{\ell=1}^{L} \frac{1}{T - \ell} \sum_{s=1}^{T-\ell} (X_s - \overline{X})(X_{s+\ell} - \overline{X}), \qquad (4.4)$$

where L satisfies the conditions that $L \to \infty$ and $L/\sqrt{T} \to 0$ (cf. Theorem 6.6.1). Thus, the standard confidence interval is based on the approximate pivotal

$$G_1(T) = \frac{\sqrt{T}(\overline{X} - \mu)}{\hat{\sigma}_T}.$$

To construct *EF based* c.i., we consider

$$\hat{\sigma}_T^2(1) = \frac{1}{T} \sum_t (X_t - \mu)^2 + 2 \sum_{\ell=1}^{L} \frac{1}{T - \ell} \sum_{s=1}^{T-\ell} (X_s - \mu)(X_{s+\ell} - \mu). \qquad (4.5)$$

Thus, the EF pivotal is given by

$$G_2(T) = \frac{\sqrt{T}(\overline{X} - \mu)}{\hat{\sigma}_T(1)},$$

which has a large sample standard normal distribution. We construct a large sample confidence interval for the unknown parameter μ, by solving the quadratic $G_2(T)^2 = Z_{\alpha/2}^2$ to get the upper and lower limits of the confidence interval. In very large samples, difference between the two pivotals may be negligible. However, for moderate sample sizes, this need not be the case.

We now report a simulation study, where $T = 60$. We take $L = 4$. In the first case, the error distribution is standard normal and $\rho = 0.6$. The true mean of the series is 0. The proportions of times the Estimator based c.i. and the EF based c.i. included the true mean are respectively 0.8785 and 0.967 respectively. The same proportions when the error distribution is that of $Y - 1$, where Y an exponential random variable with mean 1 are 0.8655 and 0.9580.

In general, with $\bar{H}(\theta) = \frac{1}{T} \sum_t H_t(\theta)$, we have the asymptotic pivotal

$$G(T) = \frac{\sqrt{T}\,\bar{H}(\theta)}{\tilde{\sigma}_T},$$

where

$$\tilde{\sigma}_T^2 = \frac{1}{T}\sum_{s=1}^{T} H_t(\theta)^2 + 2\sum_{\ell=1}^{L}\frac{1}{T-\ell}\sum_{s=1}^{T-\ell} H_s(\theta)H_{s+\ell}(\theta).$$

In general, the equations $G(T) = \pm Z_{\alpha/2}$ needs be solved by standard numerical recipes. Starting values for such iterative procedures can be taken as the end points of the Estimator based c.i..

The above approach does not require any additional assumption such as differentiability of H_t etc. For example, the confidence interval for θ, the median of X_1, follows in the same manner and is given by taking an elementary estimating function as $H_t(\theta) = I[X_t \leq \theta] - 0.5$. This discussion applies to the sample percentiles also. The estimating functions H_t, $t = 1, 2, \cdots, T$ do not satisfy regularity conditions. However, the pivotal $G(T)$ can still be constructed. We notice that such a construction avoids not only estimation of the population median, but it also avoids estimation of the p.d.f of X_1 at the population median.

References

Abdullah, N.A., Mohamed, I., Peiris, S., Azizan, N.A.: A new iterative procedure for estimation of RCA parameters based on estimating functions. Appl. Math. Sci. **5**, 193–202 (2011)

Basawa, I.V., Godambe, V.P., Taylor, R.L. (eds.) Selected Proceedings of the Symposium on Estimating Functions. Lecture Notes-Monograph Series 32. Institute of Mathematical, Statistics (1997)

Bhapkar, V.P.: On a measure of efficiency in an estimating equation Sankhyā A **34**, 467–472 (1972)

Chandrasekhar, B., Kale, B.K.: Unbiased statistical estimating functions in the presence of nuisance parameters. J. Stat. Plan. Inf. **9**, 45–54 (1984)

Chatterjee, S., Bose, A.: Generalized bootstrap for estimating equations. Ann. Stat. **33**, 414–436 (2005)

Cliff, A.D., Ord, J.K.: Spatial Processes and Applications. Pion Limited, London (1981)

Durbin, J.: Estimation of parameters in time-series regression models. J. Roy. Statist. Soc. Ser. B **22**, 139–153 (1960)

Godambe, V.P., Heyde, C.C.: Quasi-likelihood and optimal estimation. Int. Statist. Rev. **55**, 231–244 (1987)

Godambe, V.P., Thompson, M.E.: An extension of Quasi-Likelihood estimation (with discussion). J. Statist. Plan. Inf. **22**, 132–152 (1989)

Godambe, V.P.: An optimum property of regular maximum likelihood estimation Ann. Mathe. Statist. **31**, 1208–1211 (1960)

Godambe, V.P.: The foundation of finite sample estimation in stochastic processes. Biometrika **72**, 419–428 (1985)

Godambe, V.P. (ed.): Estimating Functions. Oxford University Press, Oxford (1991)

Godambe, V.P., Kale, B.K.: Estimating equations: an overview. In: Godambe, V.P. (ed.) Estimating Functions, pp. 3–20. Oxford University Press, Oxford (1991)

Hall, P., Heyde, C.C.: Martingale Limit Theory and its Applications. Academic Press, London (1980)

Hwang, S.Y., Basawa, I.V.: Godambe estimating functions and asymptotic optimal inference. Stat. Probab. Lett. **81**, 1121–1127 (2011)

Kale, B.K.: An extension of Cramér-Rao inequality for statistical estimation functions. Skand. Aktuar. **45**, 80–89 (1962)

Klimko, L.A., Nelson, P.I.: On conditional least squares estimation for stochastic processes. Ann. Statist. **6**, 629–642 (1978)

Lele, S.R.: Estimating functions in chaotic systems. J. Amer. Statist. Assoc. **89**, 512–516 (1994)

McLeish, D.L.: Estimation for aggregate models: the aggregate Markov chain Canad. J. Statist. **12**, 265–282 (1984)

McLeish, D.L., Small, C.G.: The Theory and Applications of Statistical Inference Functions Lecture notes in Statistics, vol. 44, Springer-Verlag, New York (1988)

Naik-Nimbalkar, U.V.: Estimating functions in stochastic processes. In: Prakasa Rao, B.L.S., Bhat, B.R. (eds.) Stochastic Processes and Statistical Inference, pp. 52–72. New Age International Publishers, New Delhi (1996)

Ord, J.K.: Estimation methods for models of spatial interaction. J. Amer. Statist. Assoc. **70**, 120–126 (1975)

Rao, C.R.: Linear Statistical Inference and its Applications. Wiley, New York (1965)

Whittle, P.: On stationary processes in the plane. Biometrika **41**, 434–449 (1954)

Chapter 5
Estimation of Joint Densities and Conditional Expectation

Abstract This chapter deals with estimation of joint density and conditional expectation when observations form a stationary time series. We describe kernel density estimation under the assumption that the time series satisfies the strong mixing condition. We give examples wherein kernel density estimation has been applied to real life data.

5.1 Introduction

We assume that $\{X_t, t \geq 1\}$ is a strictly stationary and real valued sequence such that, for every m, the joint distribution of $(X_1, X_2, \ldots, X_m)$ is absolutely continuous with the joint p.d.f $f(x_1, x_2, \ldots, x_m)$.

Let $k_m(\mathbf{x})$ be a bounded p.d.f. of m variables, i.e., $\int_{\Re^m} k_m(\mathbf{x})d\mathbf{x} = 1$. In what follows, the number of variables m in $\mathbf{x}$ may not be specified. Let T denote the sample size and let $A(T, m)$ denote an $m \times m$ positive definite matrix. In the sequel, we simply write A_m for $A(T, m)$. Let $\mid B \mid$ denote the determinant of a matrix B. It is assumed that each element of A_m converges to 0 as $T \to \infty$.
We write

$$K(A_m, \mathbf{x}) = \frac{k_m(A_m^{-1}\mathbf{x})}{T \mid A_m \mid}, \quad \mathbf{x} = (x_1, x_2, \ldots, x_m)'.$$

Functions such as $k_m(\mathbf{x})$ are known as kernels. In applications, it is convenient to choose $k_m(\mathbf{x})$ as the product of m p.d.f.s, i.e., $k_m(\mathbf{x}) = \prod_{i=1}^{m}(1/a_i)k_1(x_i/a_i)$ where a_i's are appropriate positive constants. In this case, $A_m = \text{diag}(a_1, a_2, \ldots, a_m)$. We list below some important kernels in one dimension.

1. $k(u) = 1/(2C), \quad \mid u \mid < C$.
2. $k(u) = 1 - \mid u \mid, \quad \mid u \mid < 1$.
3. $k(u) = (1/\sqrt{2\pi})\exp(-u^2/2), \quad -\infty < u < \infty$.
4. $k(u) = 3/(4\lambda^3)(\lambda^2 - u^2), \quad u^2 < \lambda^2, \quad \lambda > 0$.

M. B. Rajarshi, *Statistical Inference for Discrete Time Stochastic Processes*,
SpringerBriefs in Statistics, DOI: 10.1007/978-81-322-0763-4_5,
© The Author(s) 2012

Let $Y_t(m) = (X_t, X_{t+1}, \ldots, X_{t+m-1})'$ be the column random vector of m consecutive observations starting with X_t. As in the case of i.i.d. random variables, a kernel estimator of $f(x_1, x_2, \ldots, x_m)$ is given by

$$\hat{f}(x_1, x_2, \ldots, x_m) = \frac{1}{(T-m)\mid A_m \mid} \sum_{t=1}^{T-m+1} K\left(A_m, (\mathbf{x} - Y_t(m))\right).$$

For $r < m$, the estimator of $f(x_1, x_2, \ldots, x_r)$, the joint p.d.f. of $(X_1, X_2, \ldots, X_r)$ at $(x_1, x_2, \ldots, x_r)$ can be similarly defined. The estimator of the conditional p.d.f. of $(X_{r+1}, X_{r+2}, \ldots, X_m)$ at $(x_{r+1}, x_{r+2}, \ldots, x_m)$ given $X_1 = x_1, X_2 = x_2, \ldots, X_r = x_r$ is then given by

$$\hat{f}(x_{r+1}, x_{r+2}, \ldots, x_m \mid x_1, x_2, \ldots, x_r) = \hat{f}(x_1, x_2, \ldots, x_m)/\hat{f}(x_1, x_2, \ldots, x_r).$$

In particular,

$$\hat{f}(x_2 \mid x_1) = \left[\frac{TA_1}{(T-1)|A_2|}\right]\frac{\sum_{t=1}^{T-1} k_2(A_2^{-1}[(x_1, x_2)' - Y_t(2)])}{\sum_{t=1}^{T} k_1(A_1^{-1}[x_1 - X_t])}.$$

The above is of particular interest when $\{X_t, t = 1, 2, \ldots\}$ is a stationary and ergodic Markov sequence of order 1.

Though the above can be used to estimate the conditional expectation of a function of $X_{r+1}, X_{r+2}, \ldots, X_m$ given $X_1 = x_1, X_2 = x_2, \ldots, X_r = x_r$, the following approach is direct and operationally more convenient.

Let G be a function of $X_{r+1}, X_{r+2}, \ldots, X_m$. We wish to estimate

$$H(G, x_1, x_2, \ldots, x_r)$$
$$= E[G(X_{r+1}, X_{r+2}, \ldots, X_m) \mid X_1 = x_1, X_2 = x_2, \ldots, X_r = x_r].$$

Consider the statistic

$$\hat{G}_1 = (T - r - m)^{-1} \sum_{t=1}^{T-m-r} G(X_{t+r}, X_{t+r+1}, \ldots, X_{t+r+m})$$
$$K(A_r, (x_1, x_2, \ldots, x_r)' - Y_t(r)).$$

Then, the required estimator is given by

$$\hat{H}(G, x_1, x_2, \ldots, x_r) = \hat{G}_1/\hat{f}(x_1, x_2, \ldots, x_r).$$

The following cases are of interest.

(i) $m = r + 1$, $G(X_{r+1}) = I[X_{r+1} \le x]$, x fixed. This corresponds to the conditional distribution function of X_{r+1} given $X_r = x$.

(ii) $m = r+1$, $G(X_{r+1}) = X_{r+1}$. In this case, we get an estimator of the conditional expectation of X_{r+1} given $X_1 = x_1$, $X_2 = x_2, \ldots, X_r = x_r$.

It may be recalled here that the conditional expectation is the best predictor of X_{r+1} in terms of $X_1, X_2, \ldots, X_r$ in the class of unbiased predictors with finite variance. Similar estimator of the predictor of X_{r+n} can be constructed. An estimator of the conditional variance of X_{r+1} can be also obtained. This gives an estimator of the forecasting mean squared error (FMSE) of the estimator of the conditional expectation of X_{r+1}.

5.2 Main Results

Roussas (1968, 1969a,b) and Rosenblatt (1970) are some of the earliest works on kernel-based density estimation for dependent random variables. Roussas (1969b) assumes that the underlying sequence is a Markov sequence which satisfies Doeblin's condition and obtains kernel estimates of the transition density. Here, we discuss results due to Robinson (1983) and Bosq (1996). References to earlier work can be obtained from Basawa and Prakasa Rao (1980) and Bosq (1996).

Definition 5.2.1 A function $h(y)$ of m variables $(y_1, y_2, \ldots, y_m)$ is said to belong to the class $C_m(x, \lambda)$ (where $x = (x_1, x_2, \ldots, x_m)$ and λ is a positive constant), if for some $\delta > 0$, there exists a constant C such that if $\| y \| = \left(\sum_{i=1}^{m} y_i^2 \right)^{1/2} < \delta$, we have

$$| h(x - y) - P_0 - P_1 \cdots - P_r | \leq C \| y \|,$$

where $P_0 = h(x)$ and P_j for ($j \geq 1$) is a polynomial of degree j in $(y_1, y_2, \ldots, y_m)$ and r is an integer such that $r < \lambda$.

We observe that if h has continuous partial derivatives of order $r + 1$ at x, $h(y) \in C_m(x, r + 1)$.

Let $B = ((b_{ij}))$ be a matrix of order $p \times p$. Then, $\| B \| = \left(\sum_i \sum_j b_{ij}^2 \right)^{1/2}$. Robinson (1983) assumes the following.

A1. The stationary sequence $\{X_t, t \geq 1\}$ is strong mixing with $\sum_t t\alpha(t) < \infty$.

A2. The random stationary sequence $\{X_t, t \geq 1\}$ is strong mixing such that for an appropriate $\nu > 0$, $\sum_t [\alpha(t)]^{\nu/(2+\nu)} < \infty$.

Theorem 5.2.1 (Robinson 1983) *Suppose that the following assumptions hold.*

1. *The assumption A1 holds.*
2. $T \| A_m \|^{2\lambda} | A_m | \to 0$ *as* $T \to \infty$.
3. *The joint density f of m consecutive observations satisfies that $f(y) \in C_m(x, \lambda)$.*
4. $\int x_1^{h_1} x_2^{h_2} \cdots x_m^{h_m} k_m(x) \mathrm{d}x = 0$ *for all non-negative integers $h_1, h_2, \ldots, h_m$ such that $0 < h_1 + h_2 \cdots + h_m \leq s$ where s is the greatest integer less than λ.*

5. *The kernel $k_m(x)$ is bounded with a compact support*

$| k_m(x) | \leq C \exp(-D \| u \|^\rho)$ *for some positive and real C, D such that $\rho > 0$*

and $\liminf_{T \to \infty} \| A_m \|^\rho \log | A(T, m) | > -\infty$

$| k_m(x) | \leq C (1 + \| u \|)^{-m-\omega}$ *and* $\| A_m \|^{m+\omega-\lambda} \leq C | A_m |$, $\omega > \lambda$.

Then, as $T \to \infty$,

$$[T|A(T, m)|]^{1/2} \left(\hat{f}(x_1, x_2, \ldots, x_m) - f(x_1, x_2, \ldots, x_m) \right)$$

$$\xrightarrow{\mathscr{L}} N \left(0, f(x_1, x_2, \ldots, x_m) \int_{\Re^m} k_m^2(x) \mathrm{d}x \right).$$

The theorem shows that an optimal kernel is the one which minimizes $\int_{\Re^m} k_m^2(x)\mathrm{d}x$. The estimator of the conditional density suggested earlier as ratio of estimators of the marginal p.d.f.s can be easily shown to be consistent. However, it is not possible to establish its asymptotic normality. In the next theorem, we state asymptotic normality of $\hat{H}(G, x_1, x_2, \ldots, x_r)$.

Theorem 5.2.2 (Robinson 1983) *Suppose that the following assumptions hold.*

1. *The function G is bounded and the Assumption A1 holds $E(G^\gamma) < \infty$, $\gamma > 2$ and the Assumption A2 holds.*
2. *$f(x_1, x_2, \ldots, x_r) > 0$.*
3. *The function $E\left[G^2 \mid X_1 = x_1, X_2 = x_2, \ldots, X_r = x_r\right]$ is a continuous function of $x_1, x_2, \ldots, x_r$.*
4. *$E\left[G \mid X_1 = y_1, X_2 = y_2, \ldots, X_r = y_r\right] \in C_r((x_1, x_2, \ldots, x_r), \lambda)$.*
5. *$\sup_{\|y-x\|<\delta} E\left[G^{\gamma(1)} \mid X_1 = y_1, X_2 = y_2, \ldots, X_r = y_r\right] < \infty$ for $\gamma(1) > \gamma$.*
6. *The estimation of $f(x_1, x_2, \ldots, x_r)$ satisfies all the requirements of Theorem 5.1.1.*

Then,

$$[T \mid A_m \|]^{1/2}(\hat{G} - G) \xrightarrow{\mathscr{L}} N(0, V),$$

where $V = (E[G^2(x_1, x_2, \ldots, x_r)] - (E[G(x_1, x_2, \ldots, x_r)])^2)/f(x_1, x_2, \ldots, x_r)$.

We note that variances of the asymptotic normal distributions in the above two theorems can be estimated by consistent estimators . This allows one to carry out the usual statistical tests and to construct confidence intervals for such functions.

Remark 5.2.1 It is well known that performance of kernel based estimation is not so sensitive to a kernel. It is more affected by the choice of bandwidth. Further, it is more convenient to choose a kernel in m dimensions as the product of m one-dimensional kernels. While obtaining a density estimator, it is advisable to allow the bandwidth to vary with x, the point at which a density estimator is being computed. It may be pointed out that dependence among observations over a time series has no effect on the asymptotic variance, as the above results show. In particular, cross-validation methods may be attempted to select the optimal bandwidth.

Remark 5.2.2 The kernel $k_m(x)$ need not be a p.d.f. over $\mathfrak{R}^m$ and the above proof continues to be valid in such a case. In the case of i.i.d. observations, it is known that there exist kernels, which take negative values and which are not necessarily probability density functions and yet have smaller mean squared error cf. Silverman (1986). This applied to the density estimation in the case of dependent observations also.

Bosq (1996) discusses estimation of marginal p.d.f. and conditional expectation when X_t is a d-dimensional random vector under the assumption that the given series satisfies the following.

$$\sup_{\substack{A \in \sigma\{X_s\} \\ B \in \sigma\{X_{s+t}\}}} | P(A \cap B) - P(A)P(B) | \le Ct^{-\beta},$$

for C and β, both being positive. It should be noted that the above definition is in terms of σ-fields of random vectors X_s and X_{s+t} only. We notice that under the assumption that $\{X_t, t \ge 1\}$ is a Markov sequence, the above is equivalent to strong mixing. Bosq (1996) calls such sequence as 2-α-mixing.
Bosq (1996) assumes the following.

1. The joint p.d.f. f of (X_s, X_t) is continuously differentiable upto order 2. Moreover,
   $\sup_x f_{X_t}(x) < \infty$ and $\sup_{x,y} f_{X_s, X_t}(x, y) < \infty$.
2. The process is 2-α-mixing.
3. Let $g((s, t), (x, y)) = f_{X_s, X_t}(x, y) - f_{X_s}(x)f_{X_t}(y)$. Then,

$$\sup_{|t-s|\ge 1} \left\{ \int \| g((s, t), (x, y)) \|^p \, dxdy \right\}^{1/p} < \infty,$$

for some $p > 2$. Further $\beta > 2(p - 1)/(2 - p)$
                                    or
$| g((s, t), (x, y)) - g((s, t), (x', y')) | \le C \| (x, y) - (x', y') \|$. Further,
$\beta > (2d + 1)/(d + 1)$.

Let

$$\hat{f}_T(x) = \frac{1}{Ta_T^d} \sum_{t=1}^{T} K\left((x - X_t)/a_T\right)$$

be a kernel estimator of the density of $f_{X_t}(x)$, where $K(\cdot)$ is a function satisfying,

$$\lim_{\|u\|\to\infty} \| u \|^d K(u) = 0 \quad \text{and} \quad \int_{\mathfrak{R}^d} \| u \|^2 K(u) < \infty.$$

Further, $a_T \to 0$ and $T a_T^d \to \infty$. Let $a_T = C_T T^{-1/(d+4)}$, where $C_T \to C > 0$. Under these assumptions, Bosq (1996) derives a lower bound for the MSE of a density estimator and shows that a kernel-based estimator attains the optimal bound.

Bosq (1996) further discusses the case when $\{X_t, t \geq 1\}$ is GSM. He then shows that the kernel estimator is uniformly strongly consistent. Further, asymptotic normality of the kernel estimator is proved by proving a Central Limit Theorem for a triangular array of α-mixing sequences, i.e., Lindeberg-Feller type CLT . Similar results have been proved for the above kernel-based estimator of the conditional expectation.

Nze and Doukhan (2002) (see Sects. 5 and 6 of their paper) prove CAN property of the kernel-based estimators of a marginal p.d.f., under the assumption that the underlying sequence satisfies a type of weak dependence condition. They also have a Lindeberg-Feller type CLT for the weakly dependent processes, as defined therein. It needs to be pointed out that some of the nonlinear time series models discussed in Tjøstheim (1994), (cf. Example 1.3.7) do not satisfy the smoothness conditions as required in our discussion above.

Secondly, functional estimation can be inefficient particularly for higher dimensional data and if a parametric model gives a good fit to the data, estimation and forecasting procedures based on such a parametric model are superior to the functional estimation.

Methods other than those based on kernel have been discussed in the literature. Yakowitz (1989) proves consistency of density estimators without assuming any mixing conditions. He assumes that the process is Markovian and the p.d.f. is estimated based on nearest neighbor methods, cf. Prakasa Rao (1996).

For a review of functional estimation in stochastic models, we refer to Prakasa Rao (1996). Basawa and Prakasa Rao (1980), (Chap. 11) contains a discussion of earlier results on kernel density estimation obtained under the assumptions of Doeblin's conditions for Markov sequences. It is shown that such estimators are asymptotically unbiased estimators. Results on density estimation by the orthogonal series under the assumption of ϕ-mixing and the delta-sequence method for Markov sequences (Prakasa Rao 1978) have also been discussed therein.

Some examples of data analysis using density estimation
1. Robinson (1983) illustrates estimation of conditional expectation by analyzing the well-known Wolfer sunspot series. This series has patterns of nonlinearity and it does not get explained by a Gaussian distribution. Robinson picks up lag 1 and lag 9 as the sample ACF shows a strong dependence at these lags. We refer to page 194 of Robinson (1983), where the graphs of estimates of $E(X_t \mid X_{t-j}, X_{t-k})$ for $(j, k) = (1, 2), (1, 9), (2, 9)$ are given. The product kernel with Gaussian kernel as marginal densities was used. These graphs clearly point out a strong nonlinear pattern in such conditional expectations. It looks formidable to capture such patterns by parametric modeling.
2. Yakowitz (1985) applies both the Box-Jenkins ARMA techniques and estimator of the conditional expectation to obtain one-step ahead predictors of daily flows of Kooteni river (in USA), between 1911–1933. His analysis shows that the two methods agree well.

3. The Appendix of Bosq (1996) compares the non-parametric predictors and Box-Jenkins ARMA-based predictors for simulated time series data as well as data sets on profit margin, cigar consumption, changes in business inventories, coal production, French car registrations and French electricity consumption. To study efficacy of predictors $\tilde{X}_i$ of X_i of the p future random variables, he defines the performance criterion Erreur relative Moyenne Observée (EMO) as $\text{EMO} = (1/p) \sum_i |X_i - \tilde{X}_i| / |X_i|$ i.e., observed mean relative error. In a majority (12 out of 17) of the cases, the non-parametric predictor is better than the best ARMA-based predictor with respect to the EMO criterion.

References

Basawa, I.V., Prakasa Rao, B.L.S.: Statistical Inference for Stochastic Processes. Academic Press, London (1980)

Bosq, D.: Nonparametric Statistics for Stochastic Processes Lecture Notes in Statistics, vol. 110. Springer-Verlag, New York (1996)

Nze, P.A., Doukhan, P.: Weak dependence: models and applications. In: Dehling, H., Mikosch, T., Sørensen, M. (eds.) Empirical Process Techniques for Dependent Data, pp. 117–136. Birkhäuser, Boston (2002)

Prakasa Rao, B.L.S.. Density estimation for Markov processes using delta sequences. Ann. Inst. Statist. Math. **30**, 73–87 (1978)

Prakasa Rao, B.L.S.: Nonparametric approach in time series analysis. In: Prakasa Rao, B.L.S., Bhat, B.R. (eds.) Stochastic Processes and Statistical Inference, pp. 73–89. New Age International Publishers, New Delhi (1996)

Robinson, P.M.: Nonparametric estimators for time series. J. Time Ser. Anal. **4**, 185–297 (1983)

Rosenblatt, M.: Density estimation and Markov processes. In: Puri, M.L. (ed.) Nonparametric Techniques in Statistical Inference, pp. 199–210. Cambridge University Press, Cambridge (1970)

Roussas, G.G.: On some properties of nonparametric estimates of probability density functions. Bull. Soc. Mathematique de Greece **2**, 29–43 (1968)

Roussas, G.G.: Nonparametric estimation in Markov processes. Ann. Inst. Statist. Math. **21**, 73–87 (1969a)

Roussas, G.G.: Nonparametric estimation of the transition density function. Ann. Mathe. Statist. **40**, 1386–1400 (1969b)

Silverman, B.W.: Density Estimation for Statistics and Data Analysis. Chapman and Hall, London (1986)

Tjøstheim, D.: Non-linear time series: a selective review scand. J. Statist. **21**, 97–130 (1994)

Yakowitz, S.: Markov flow models and the flood warning problem. Water Resour. Res. **21**, 81–88 (1985)

Yakowitz, S.: Nonparametric density and regression estimation for Markov sequences without mixing assumptions. J. Multivar. Anal. **30**, 124–136 (1989)

Chapter 6
Bootstrap and Other Resampling Procedures

Abstract In this chapter, we discuss various resampling procedures, such as bootstrap, jackknife, and sample re-use procedures for discrete time stochastic processes. Our discussion begins with bootstrap procedures for finite and infinite Markov chains. Further, bootstrap for stationary real valued Markov sequences based on transition density estimators is discussed. This is followed by bootstrap based on residuals for stationary and invertible linear ARMA time series. We then describe bootstrap and jackknife procedures based on blocks of stationary observations. Further, we discuss a bootstrap procedure based on AR-sieves. Results which prove superiority of block-based bootstrap over the traditional central limit theorems are discussed herein. The last section discusses bootstrap procedures for construction of confidence intervals based on estimating functions.

6.1 Efron's Bootstrap

It is now well established that resampling procedures such as bootstrap offer easy-to-use, yet powerful methods for estimation of sampling distribution of estimators, test statistics and approximate pivotals for construction of confidence intervals. In the case of i.i.d. observations, the traditional techniques require derivations such as δ-method (Serfling 1980, Sect. 3.1), CLTs and it is often viable to do so. However, bootstrap is very easy to use and quite frequently, gives more accurate answers for estimation of the sampling distributions. Naturally, it has become an integral part of a number of statistical packages.

In the case of stochastic models, derivations such as required to implement δ-method or CLT, are quite complicated. As rightly pointed out Künsch (1989), bootstrap and other procedures are not only a boon but probably more of a necessity in statistical analysis of stochastic models. In fact, the traditional analysis of parametric models can be so cumbersome that parametric bootstrap methods are very helpful in inference of parametric stochastic models.

M. B. Rajarshi, *Statistical Inference for Discrete Time Stochastic Processes*,
SpringerBriefs in Statistics, DOI: 10.1007/978-81-322-0763-4_6,
© The Author(s) 2012

Efron's bootstrap

Let $\{X_1, X_2, \ldots, X_T\}$ be i.i.d. observations from a population with the distribution function F. Let $\theta = H(F)$ be a function of the distribution function F. Let us assume that observations are real valued and let $F_T(x) = (1/T) \sum_{t=1}^{T} I[X_t \leq x]$ be the empirical distribution function. Let H_T be a symmetric function of $\{X_1, X_2, \ldots, X_T\}$, we write $\hat{\theta}_T = H(F_T)$. A large number of estimators can be written in this form.

(a) *Estimation of variance of $\hat{\theta}_T$.*

Suppose that we are interested in estimating $\sigma_T^2 = \text{Var}(\hat{\theta})$. The bootstrap algorithm is as follows.

1. Draw a simple random sample with replacement (SRSWR) of size T from $\{X_1, X_2, \ldots, X_T\}$. As is customary, a star (∗) refers to a bootstrap observation. For example, a bootstrap sample is denoted by $\{X_1^*, X_2^*, \ldots, X_T^*\}$.
2. Estimate θ from these T bootstrap observations, exactly the same way $\hat{\theta}_T$ is computed from $\{X_1, X_2, \ldots, X_T\}$.
3. Repeat 1. and 2. above B times to get B estimators, denoted by $\hat{\theta}_T^*(1), \hat{\theta}_T^*(2)$, $\ldots, \hat{\theta}_T^*(B)$
4. Let $\hat{\sigma}_T^2(Boot)$ be the variance of B values $\hat{\theta}_T^*(1), \hat{\theta}_T^*(2), \ldots, \hat{\theta}_T^*(B)$ as obtained in
5. This is the bootstrap estimator of σ_T^2 .

The number B needs to be sufficiently large, say, at least 1,000.

(b) *Construction of bootstrap confidence interval for θ*

A bootstrap confidence interval is constructed as follows. Let $\hat{\sigma}_T^*(b)$ be an estimator based on the bth bootstrap sample (it is computed exactly the same way $\hat{\sigma}_T$ is computed from the sample that we have actually observed). We have B values of $S_T^*(b) = (\hat{\theta}_T^*(b) - \hat{\theta}_T)/\hat{\sigma}_T^*(b), b = 1, 2, \ldots, B$. The $[100(\alpha/2)]$th and $[100(1 - \alpha/2)]$th order statistics of these B values are bootstrap estimates of the corresponding percentiles of the asymptotic distribution of $S_T = (\hat{\theta}_T - \theta)/\hat{\sigma}_T$. (We note that $T\hat{\sigma}_T^2$ estimates the variance of the asymptotic normal distribution of $\sqrt{T}(\hat{\theta}_T - \theta)$). We use these percentiles as the "table values", which replace the corresponding percentiles of the standard normal distribution. The rest of the construction of a confidence interval is similar to the standard confidence interval. We notice that for a bootstrap confidence interval, the upper and lower limits need not to be equidistant from $\hat{\theta}_T$.

Here is an explanation why the Efron's bootstrap works. To find the standard error of an estimator or to obtain a confidence interval for an unknown parameter, suppose for the time being that the underlying distribution function F is known. However, we are not in a position to carry out any theoretical derivation. For example, consider the derivation of the t-distribution of the pivotal $\sqrt{T}(\hat{X}_T - \mu)/\hat{\sigma}_T$, where μ is the mean of a normal distribution. Since we know F, we can simulate the estimator or the pivotal for a large number of times and can come up with a very good approximation to the sampling distribution or the variance of a statistic.

But, we do not know F! Instead, we have an estimator F_T of F. Thus, instead of simulating from F, we can simulate from F_T to get repeated samples. This is what Efron's bootstrap does. We note that F_T is the distribution function of a discrete random variable which assigns probability $1/T$ to each of the observations. Sampling from F_T is thus equivalent to obtaining a SRSWR of size T from the sample $X_1, X_2, \ldots, X_T$. We also note that F_T is a uniformly strongly consistent estimator of F i.e., $\sup_{x \in \Re} \mid F_T(x) - F(x) \mid \to 0$ a.s. In the case of i.i.d. observations, the distribution function F together with the i.i.d. assumption constitutes a model for the observations. This model has been correctly estimated and the estimator has been used to obtain repeated samples.

For a very useful discussion of bootstrap, we refer to Chernick (2008). Discussion in Chaps. 5, 8 and 9 of Chernick (2008) particularly refer to bootstrap methods in stochastic models. Efron and Tibshirani (1993), Shao and Tu (1995) and Davison and Hinkley (1997) offer extensive discussions of bootstrap procedures. These books also include bootstrap and its applications in time series models. Lahiri (2003) exclusively deals with resampling procedures for dependent data.

The symbols P^*, E^*, and Var*, respectively denote conditional probability of an event, conditional expectation and conditional variance of a random variable, given the sample.

It is important to record well-known results on validity as well as superiority of Efron's bootstrap in the case of i.i.d. observations. For various classes of estimators, under mild conditions, a bootstrap estimator of the variance of the asymptotic normal distribution is consistent. Further, if $\hat{\theta}$ is a continuously differentiable function of sample moments, under additional assumptions,

$$\sup_x \left| P^* \left[\frac{\hat{\theta}_T^* - \hat{\theta}_T}{\hat{\sigma}_T^*} \leq x \right] - P \left[\frac{\hat{\theta}_T - \theta}{\hat{\sigma}_T} \leq x \right] \right| = o_p \left(\frac{1}{\sqrt{T}} \right)$$

which shows that the bootstrap approximation is better than the traditional CLT approximation, which has the error rate of the order $1/(\sqrt{T})$. For results of this type, we refer to Singh (1981), Babu and singh (1984) and Hall (1992) for a comprehensive treatment. It needs to be mentioned that without the above Studentization, the bootstrap gives an approximation which is of the same error rate as that of CLT.

In the sequel, we describe a bootstrap procedure, i.e., a procedure to generate a bootstrap sample path. The remaining part of the bootstrap methodology is the same as above, unless a deviation is specified.

6.2 Markov Chains

Let $(X_0, X_1, X_2, \ldots, X_T)$ be a realization from a first order finite Markov chain with P as its one-step t.p.m. We assume that the Markov chain is irreducible and aperiodic. From Sect. 2.1, we recall that $\hat{p}_{ij} = N_{ij} / \sum_j N_{ij}$, if $\sum_j N_{ij} > 0$ and

$\hat{p}_{ij} = 0$ if $\sum_j N_{ij} = 0$. Thus, $\hat{P} = ((\hat{p}_{ij}))$ is the non-parametric m.l.e. of P. A suitable estimate of the initial distribution, i.e., the distribution of X_0, $\hat{P}$ and the assumption of first order Markov property constitute estimate of the statistical model for $\{X_t, t = 0, 1, 2, \ldots\}$. If it can be assumed that the Markov chain is stationary, the initial distribution is estimated by $\hat{\pi}$, the stationary distribution of the chain with the t.p.m. $\hat{P}$. In this case, the distribution of X_0^*, the initial value of a bootstrap sample path, is given by $\hat{\pi}$. If the chain cannot be assumed to be in equilibrium, all bootstrap sample path may start with X_0. Further, $P^*[X^*(t + 1) = j | X^*(t) = i, X^* (t-1), \ldots, X^*(0)] = \hat{p}_{ij}$, for all t, (i, j) and all past values of $X^*(t-1), \ldots, X^*(0)$. Under the assumptions on the Markov chain, for a moderately large T, the effect of the initial distribution is negligible.

Kulperger and Prakasa Rao (1989) justify the above bootstrap procedure as follows. Let P and Q be $M \times M$ t.p.m.s and let $\| P - Q \| = \max_{ij} | p_{ij} - q_{ij} |$ be the distance function on the space of $M \times M$ t.p.m.s. Let $H(P)$ be a real valued continuous function of P with respect to the metric $\| P - Q \|$. We have the following theorem.

Theorem 6.2.1 (Kulperger and Prakasa Rao 1989) *Assume that the limiting distribution of $\sqrt{T}(H(\hat{P}) - H(P))$ exists. Let E be a P-continuous set, i.e., the measure of the boundary set of E with respect to the limiting distribution is 0, when P is the true one-step t.p.m. Suppose that $\| P - Q \| < \delta$, so that for a suitable δ, the Markov chain with Q as the t.p.m. is irreducible and ergodic. Let $[\mathcal{L}(\sqrt{T}(H(\hat{P}) - H(P))), E; P]$ denote the probability of the set E under the limiting law of $\sqrt{T}(H(\hat{P}) - H(P))$ when the underlying Markov chain follows the t.p.m. P. Then, under the model P,*

$$\lim_{T \to \infty} \lim_{\delta \to 0} |[\mathcal{L}(\sqrt{T}(H(\hat{P}) - H(P))), E; P]$$

$$-[\mathcal{L}(\sqrt{T}(H(\hat{Q}) - H(Q))), E; Q]| = 0.$$

Validity of bootstrap for such functions of P follows since $\| \hat{P} - P \| \to 0$, a.s.(P). An estimator of a real valued function H of the stationary distribution π of P, can be bootstrapped. Kulperger and Prakasa Rao (1989) illustrate their results by the following functional. Let $W = \inf\{t \mid X_t = j\}$. Then, $H(P) = E[W \mid X_0 = i]$ can be shown to be a continuous function of P. We notice that a sizable work will be required to get the asymptotic distribution of $H(\hat{P})$, in fact, the estimator of $H(P)$ itself involves a set of matrix operations. Bootstrap is very handy in this case.

Basawa Mallik McCormick and Taylor (1990) also prove validity of the above bootstrap for finite Markov chains. They suggest another bootstrap which may be described as a conditional bootstrap . In the conditional bootstrap, frequencies of M states are held fixed and multinomial samples are drawn where, the number of trials is the frequency of a state and the corresponding row of the $\hat{P}$ is the vector of multinomial probabilities. This bootstrap is based on the similarity of the likelihood of finite Markov chains (cf. Sect. 2.1) with likelihood of M independent multinomials.

Infinite Markov chains

Now, let us assume that $\{X_t, t = 0, 1, 2, \ldots\}$ is a first order Markov chain with a countably infinite state-space. It is assumed that the Markov chain is non-null persistent, aperiodic, and irreducible. A bootstrap sample path is generated in a manner similar to that of a finite Markov chain, described above. Validity of bootstrap in this case has been established by Athreya and Fuh (1992a,b). It is possible that $\hat{P}_T$, the m.l.e. of P, the one-step transition probability matrix, does not correspond to an irreducible and aperiodic Markov chain. The matrix $\hat{P}_T$ can be then modified so that the corresponding Markov chain is irreducible and aperiodic. Results on validity of bootstrap are based on laws of large numbers and central limit theorems for a double array of Markov chains, developed by Athreya and Fuh (1992a,b). They prove that the asymptotic conditional distribution of $\sqrt{T}(\hat{P}_T^* - \hat{P}_T)$ is the same as that of $\sqrt{T}(\hat{P}_T - P)$, for almost all sample paths. The weak convergence here means that a finite dimensional distribution in each case converges weakly to the same random variable.

Athreya and Fuh (1992a,b) also suggest another method of bootstrapping Markov chains, which is based on the property of irreducible non-null persistent Markov chains, that a state i is visited infinitely often (a.s.) and that times between two such returns form an i.i.d. sequence. Let $N(i)$ be the number of times the chain returns to the state i. In view of the ergodic properties, $N(i) \to \infty$ *a.s.*. We note that for such a bootstrap, $N(i)$ corresponds to a sample size in the i.i.d. case. (One should choose the state i which occurs with maximum frequency). Such a bootstrap is known as "recurrent" bootstrap. An estimator of a proportion of an event is constructed based on the histories in $N(i)$ cycles. A bootstrap estimator is similarly computed. The recurrent bootstrap is particularly useful for estimation of the sampling distribution of the estimator of $P[W_{k\ell} \leq x]$ where k and ℓ are two states and $W_{k\ell}$ is the waiting time to reach the state ℓ, starting from the state k, cf. Sect. 2 of Athreya and Fuh (1992a,b).

6.3 Markov Sequences

Let $\{X_t, t = 1, 2, \ldots\}$ be a first order Markov sequence with $\Re$ as the state-space. A kernel-based estimator of the transition density is given by $\hat{f}(y|x)$, cf. Chap. 5. Rajarshi (1990) suggests, given that $X_t^* = x_t^*$, a bootstrap observation X_{t+1}^* is generated from the density $\hat{f}(x_{t+1}|x_t^*)$, so that bootstrap observations follow the Markov property. However, Paparoditis and Politis (2002) suggest a simpler procedure which does not explicitly need an estimate of the transition density of the underlying Markov sequence. Their assumptions are as follows.

A1. Let $Y_t = (X_t, X_{t-1}, \ldots, X_{t-p+1})$. The stochastic process $\{Y_t, t \geq p\}$ forms an aperiodic, strictly stationary, and geometrically ergodic Markov sequence on the state-space $\Re^p$. Consequently, $\{X_t, t = 1, 2, \ldots\}$ is aperiodic, strictly stationary, and geometrically ergodic Markov sequence. Let $F(y|x)$ denote the one-step transition distribution function of $\{X_t, t \geq 1\}$, i.e.,

$$F(v|u) = P[X_{t+1} \le v | X_t = u] \text{ for } u, \ v \in \Re.$$

Further, let

$$F_Y(y|x) = P[Y_{t+1} \le y | Y_t = x] \text{ for } x, \ y \in \Re^p$$

be the one-step transition distribution function of the Markov sequence $\{Y_t, t \ge p\}$. (As before, the vector notation $Y \le y$, where $Y = (Y_1, Y_2, \ldots, Y_p)$, $y = (y_1, y_2, y_p)$ is to be interpreted as $\cap_j Y_j \le y_j$). The transition distribution function $F(v|u)$, $u, \ v \in \Re$ uniquely characterizes the law of the stationary process $\{X_t, t \ge 1\}$.

A2. The distribution functions $F(y)$, $y \in \Re^p$ of Y_t and $F(x|y) = P[X_{t+1} \le x \mid Y_t = y]$, $x \in S$ are absolutely continuous with respect to the Lebesgue measures on $\Re^p$ and S, respectively.

A3. For all $u \in \Re \cup \infty$,

$$\left| \int_{-\infty}^{u} f_{Y_t, X_{t+1}}(y_1, x) \mathrm{d}x - \int_{-\infty}^{u} f_{Y_t, X_{t+1}}(y_2, x) \mathrm{d}x \right| \le L(u) \parallel y_1 - y_2 \parallel,$$

where $\sup L(u) < \infty$ and $\inf L(u) > 0$.

A4. Let $f(x|y)$ denote the conditional p.d.f. of X_{t+1} at x given $Y_t = y$. Then, for $x_1, \ x_2 \in S$,

$$|f(x_1|y) - f(v_2|y)| \le C(y) \mid x_1 - x_2 \mid,$$

where $\sup C(y) < \infty$. Moreover, there is a compact set A of $\Re$ such that $X_t \in A$, a.s. Further, $f(x|y) > 0 \ \forall x \in S$.

Assumptions on the kernel K used in generating a bootstrap sequence are described below. Let K be a probability density on R^p. As in the case of density estimation, the kernel K satisfies the following.

B1. K is a bounded and symmetric probability density on R^p satisfying $K(y) > 0$ for all y. Further, $\int y K(y) \mathrm{d}y = 0$ and $\int \parallel y \parallel K(y) \mathrm{d}y < \infty$. Moreover, the kernel K is first order Lipschitz continuous, i.e., there exists a constant C such that $\parallel K(z) - K(y) \parallel \le C \parallel z - y \parallel$.

B2. There exist constants c_1, c_2 such that $0 < c_1 \le bT^\delta \le c_2$ for $0 < \delta < 1/(2p)$. We set following $K_c(y) = c^{-p} K(y/c)$ for all $c > 0$. Let $b(= b_T)$ be the bandwidth sequence.

A bootstrap sample path $(X_1^*, X_2^*, \ldots, X_T^*)$ is generated as follows.

1. We set $(X_1^*, X_2^*, \ldots, X_p^*) = (X_1, X_2, \ldots X_p)$.
2. Suppose we have generated $(X_1^*, X_2^*, \ldots, X_t^*)$. Let $Y_t^* = (X_t^*, X_{t-1}^*, X_{t-2}^*, \ldots, X_{t-p+1}^*)$, for $t = (p-1), p, \ldots, T$. Let $N_{(p,T-1)} = \{p, p+1, \ldots, T-1\}$.

Then, a discrete random variable J is defined by

$$\tilde{P}[J = j] = \frac{K_b(Y_t^* - Y_j)}{\sum\limits_{r \in N_{(p,T-1)}} K_b(Y_t^* - Y_r)}, \quad j \in N_{(p,T-1)}.$$

If $J = j$, X_{t+1}^*, the next observation in a bootstrap Markov sequence is given by X_{j+1}. This bootstrap is known as a local bootstrap procedure. Like the Efron's bootstrap for the i.i.d. observations, the above bootstrap generates as a discrete state-space sequence (in fact, a Markov chain). This property does not hold for procedure suggested in Rajarshi (1990).

Paparoditis and Politis (2002) show that, if the kernel K satisfies the assumption B1 and B2 (for every $b > 0$), there exists a t_0 such that, with probability 1, given the observations, the bootstrap sequence $\{Y_t^*, t \geq t_0\}$ is a positive recurrent, irreducible and aperiodic Markov chain with at most countable state-space. Further, the bootstrap sequence $\{X_t^*, t \geq t_0\}$ is conditionally a p-th order Markov chain. Let

$$\hat{F}(x|y) = P^*[X_{t+1}^* \leq x | Y_t^* = y], \quad x \in S, \ y \in \Re^p$$

be the conditional distribution function of X_{t+1}^* given Y_t^*. The main theorem is as follows.

Theorem 6.3.1 (Paparoditis and Politis (2002)) *Under the assumptions A1–A3, B1 and B2,*

$$\sup_{x \in S, \ y \in \Re^p} |\hat{F}(x|y) - F(x|y)| \to 0 \quad a.s.$$

In fact, any bootstrap procedure for which an estimator of $F(x|y)$ has the above property yields asymptotically correct results. Now, let us assume that the underlying Markov sequence is ρ-mixing with a geometric rate of decay. Paparoditis and Politis (2002) show that, for a large enough T_0, the bootstrap Markov sequence is almost surely ρ-mixing with the same ρ-mixing coefficients. Then, it is proved that the bootstrap estimator of variance of a mean-like estimator is strongly consistent and that the sampling distribution of such an estimator is consistently estimated by the bootstrap for large samples. Rajarshi (1990) obtains these results under the assumption that the underlying Markov sequence is ϕ-mixing .

Paparoditis and Politis (2002) give an interesting application of the above bootstrap procedure to test the null hypothesis that the underlying Markov sequence is reversible.

6.4 Bootstrap for Stationary and Invertible ARMA Series

Let $\{X_t, t \geq (1 - p)\}$ be a stationary $ARMA(p,q)$ process defined by

$$X_t = \sum_{i=1}^{p} \alpha_i X_{t-i} + \sum_{j=1}^{q} \beta_j \varepsilon_{t-j} + \varepsilon_t,$$

where (i) p and q are non-negative integers, (ii) the white noise sequence $\{\varepsilon_t\}$ is a sequence of i.i.d. random variables with $E(\varepsilon_t) = 0$ and (iii) α_i, $i = 1, 2, \ldots, p$; β_j, $j = 1, 2, \ldots, q$ are real parameters such that each of the polynomials $\alpha(z) = 1 - \sum_{i=1}^{p} \alpha_i z^i$ and $\beta(z) = 1 + \sum_{j=1}^{q} \beta_j z^j$ (where z is a complex number) does not vanish on the set $\{z| |z| \le 1\}$. Further, $\alpha(z)$ and $\beta(z)$ have no common zero. Under these assumptions, the process is causal and invertible. We further assume $\alpha_p \ne 0$ and $\beta_q \ne 0$. Define, for $|z| \le 1$, $[\beta(z)]^{-1} = \sum_{j=0}^{\infty} \gamma_j z^j$. We define $\beta_0 = \alpha_0 = 0$ and the coefficients α_j's and β_j's with negative suffixes are defined to be 0. Further, for $t \ge 1$, ε_t can be written as

$$\varepsilon_t = \sum_{k}^{t-1} \gamma_k \left\{ X_{t-k} - \sum_{i=1}^{p} \alpha_i X_{t-k-i} \right\}.$$

(cf. Brockwell and Davis (1987)). We write $\theta = (\alpha_1, \alpha_2, \ldots, \alpha_p, \beta_1, \beta_2, \ldots, \beta_q)$. Let $\{X_{1-p}, X_{2-p}, \ldots, X_0, X_1, X_2, \ldots, X_T\}$ denote the observed time series. The estimation procedures for estimating θ is as follows. Let us consider the estimating function

$$\Psi_T(\theta) = \frac{1}{\sqrt{T}} \sum_{t=1}^{T} \psi(\varepsilon_t(\theta)) Z(t-1, \theta),$$

where

$$\psi(\varepsilon_t(\theta)) = \sum_{k=1}^{t-1} \gamma_k(\theta) \left[X_{t-k} - \sum_{i=1}^{p} \alpha_i X_{t-k-i} \right]$$

and $Z(t-1, \theta)$ is a $(p+q) \times 1$ random vector defined by

$$Z(t-1, \theta) = \sum_{k=0}^{t-1} \gamma_k(\theta)[X_{t-k-1}, X_{t-k-2}, \ldots, X_{t-k-p}, \varepsilon_{t-k-1}(\theta), \ldots, \varepsilon_{t-k-q}(\theta)]'$$

with $\varepsilon_t(\theta) = 0$ if $t < 0$. In the above, the function $\psi(\cdot)$ satisfies the conditions that $E[\psi(\varepsilon_t)] = 0$, $\text{Var}[\psi(\varepsilon_t)] < \infty$ and $\psi(\varepsilon_t)$ is a twice continuously differentiable function with bounded derivatives ψ' and ψ'' (For example, for LSE, we have $\psi(x) = x$). Under the additional assumption that $E[\varepsilon_t^4] < \infty$, Kreiss and Franke (1992) prove that there is a $\sqrt{T}$-CAN solution $\hat{\theta}$ of the above estimating equation.

The following bootstrap procedure is due to Kreiss and Franke (1992) who prove validity of the bootstrap for a class of M estimators. For carrying out residual-based bootstrap, we need to recover the white noise process from the observations which will be mimicked for a given sample using estimates of the auto-regressive and moving average parameters. Thus, we have, for $1 \leq t \leq T$,

$$\tilde{r}_t = \sum_k^{t-1} \hat{\gamma}_k \left\{ X_{t-k} - \sum_{i=1}^{p} \hat{\alpha}_i X_{t-k-i} \right\}.$$

Importance of ensuring that the sum of the residuals to be 0 has been emphasized by several authors. We set

$$r_t = \tilde{r}_t - \frac{1}{T} \sum_t \tilde{r}_t,$$

so that r_t's have zero mean and can be proxies for ε_t's. A bootstrap time series is then defined as follows.

1. For $t \leq -\max(p, q)$, set $X_t^* = 0$, $r_t^* = 0$.
2. For $t > -\max(p, q)$, at each time t, select r_t^* randomly from the set $\{r_1, r_2, \ldots, r_T\}$. (This leads to a SRSWR sample from $\{r_1, r_2, \ldots, r_T\}$, the set of residuals.)
3. A bootstrap observation X_t^* is then defined recursively:

$$X_t^* = \sum_{i=1}^{p} \hat{\alpha}_i X_{t-i}^* + \sum_{j=1}^{q} \hat{\beta}_j r_{t-j}^* + r_t^*.$$

Consistency of $\hat{\theta}_T$ ensures that a bootstrap time series is stationary and invertible. Kreiss and Franke (1992) show that the above bootstrap procedure leads to a consistent estimator of the sampling distribution of $\sqrt{T}(\hat{\theta} - \theta)$ for a class of M estimators. Let

$$\Gamma_T = \frac{1}{T} \sum_t \psi'(\varepsilon_t) Z_t(j-1) Z_t(j-1)^{\mathrm{tr}},$$

where tr denotes the transpose of a matrix. It can be shown that

$$\sqrt{T}(\hat{\theta} - \theta) = \frac{1}{\sqrt{T}} [\hat{\Gamma}_T]^{-1} \Psi_T + o_p(1)$$

which leads to the asymptotic normality of $\sqrt{T}(\hat{\theta} - \theta)$. Kreiss and Franke (1992) suggest that the distribution of $\sqrt{T}(\hat{\theta} - \theta)$ be approximated by the conditional distribution of $\frac{1}{\sqrt{T}} [\hat{\Gamma}_T^*]^{-1} \Psi_T^*$, instead of following the usual procedure of approximating it by the conditional distribution of $\sqrt{T}(\hat{\theta}^* - \hat{\theta})$.

However, Allen and Datta (1999) report that this modification is helpful in MA(1) model and not very helpful in AR(1) model. Allen and Datta (1999) point out that it is more appropriate to bootstrap an appropriately studentized pivotal and prove that the residual-based bootstrap is asymptotically valid for estimation of a studentized pivotal. They assume that $E|\varepsilon_t|^3 < \infty$. Simulation results in Allen and Datta (1999) confirm that performance of such an approximation is reasonable and considerably better than the Kreiss and Franke procedure.

The following procedure is often recommended. Let $\psi_1(x) = \psi(x) - E^*[\varepsilon^*]$ and a bootstrap estimator of θ is obtained by solving

$$\Psi_T^*(\theta) = \frac{1}{\sqrt{T}} \sum_{t=1}^{T} \psi_1(\varepsilon_t(\theta)) Z(t-1, \theta).$$

Such a correction results in a (conditionally) unbiased estimating function and has been noted to be essential for validity of residual-based bootstrap procedures by many authors, see Lahiri (2003), Sect. 4.3.

Bose (1988) discusses residual-based bootstrap for AR(p) sequences. He proves that, in this case, the residual-based bootstrap outperforms the traditional central limit theorem approximation to the sampling distribution of least square estimators. His results are based on the Edgeworth expansions obtained by Götze and Hipp (1983). Bose (1990) discusses similar result for bootstrap of the LSE in MA(1) time series. It may be remarked that, a MA(1) model, the LSE has low efficiency as compared to the MLE.

6.5 AR-Sieve Bootstrap

Bickel and Bühlmann (1999) suggest an AR sieve-based bootstrap for a linear time series, though it performs well for other time series also. The AR sieve bootstrap is based on the property that a stationary and second order stationary, linear and invertible time series such as ARMA(p, q) can be represented by an infinite order AR model

$$X_t - \mu = \sum_{s=t-1}^{\infty} \phi_s(X_s - \mu) + \varepsilon_t,$$

where $\{\varepsilon_t\}$ is an i.i.d. sequence, cf. Brockwell and Davis (1987). In AR(p) sieves bootstrap procedure, we approximate the above process by a AR(p) model

$$X_t - \mu = \sum_{s=t-1}^{p} \phi_s(X_s - \mu) + \varepsilon_t.$$

For each p, we estimate μ by the sample mean $\bar{X}$ and the AR parameters ϕ's by solving the sample Yule-Walker equations. We then select p by applying the AIC assuming that the innovations have a Gaussian distribution. However, the assumption of a Gaussian distribution is only to allow us to select a data driven p. Let $r_t = X_t - \sum_{i=1}^{p} \hat{\phi}_i X_{t-i}$, $t \geq (p+1)$. Let $\bar{r}$ be their mean and $R_t = r_t - \bar{r}$. The bootstrap is carried out by selecting residuals randomly from the set of R_t's. Thus, we have $X_t^* - \hat{X} = \sum_{s=t-1}^{p} \hat{\phi}_s(X_s^* - \bar{X}) + R_t^*$.

To achieve stationarity, we set $(X^*_{-u}X^*_{-u+1}, \ldots, X^*_{-u+p+1}) = (\bar{X}, \bar{X}, \ldots, \bar{X})$ for a very large value u and then generate the successive observations to reach $(X^*_1, X^*_2, \ldots, X^*_T)$. Such a bootstrap sample would correspond to a stationary time series. Let $\hat{\theta}$ be the estimator of θ, a parameter of interest. Since, in general, $E^*(\hat{\theta}^*) \neq \hat{\theta}$, to compute $E^*(\hat{\theta}^*)$, we proceed as follows. We generate $(X^*_1, X^*_2, \ldots, X^*_N)$, where N is an integer which is very large as compared to T. The estimator of θ, say $\tilde{\theta}$, computed from such a very large sample, is taken to define the bootstrap pivotal $\sqrt{T}(\hat{\theta}^* - \tilde{\theta})$.

In Rohan and Ramanathan (2011), it is suggested to select the order of an AR model so as to minimize the mean squared error of an estimator of interest. They also assume that the underlying model is a linear ARMA series and approximate it by an AR model. The parameter of interest is the vector of auto-regressive parameters which are estimated by LSEs. It has been shown there this criterion performs well in comparison with the AIC and other information-based criteria when the underlying distribution is Gaussian and other situations also. It would be interesting to employ this technique to carry out the AR-Sieve bootstrap.

6.6 Block-Based Bootstraps for Stationary Sequences

Probability model of a stationary stochastic process, without any assumptions such as an ARMA or Markovian property, is specified by its family of finite dimensional distribution functions. In the Markov cases and in structural models, each finite dimensional distribution function can be written in terms of few parameters and finitely many distribution functions. Once these are estimated in an appropriate manner, in principle, bootstrap can be carried out based on such estimates. In a completely non-parametric situation, we assume that the sequence is of short memory, a property that is frequently characterized by the rate at which strong mixing coefficients decay. Block-based bootstrap procedures (Künsch 1989, Liu and Singh 1992) estimate the distribution of L consecutive observations by K replicates from a sample of size T. To capture the distribution of a statistic or a pivotal, L is allowed to increase at an appropriate rate.

Let $\hat{F}_m(T)$ be defined by

$$\hat{F}_m(T) = (T - m + 1)^{-1} \sum_{i=1}^{T-m+1} \delta_{X_{i+1}, X_{i+2}, \ldots, X_{i+m}},$$

where δ_y assigns probability one to the point y. The function δ_y denotes a distribution function also. Let $F(m)$ denote the distribution function of $X_1, X_2, \ldots, X_m$. An estimator $\hat{\theta}$ of the parameter $\theta(F(m))$ is then defined by

$$\hat{\theta} = \theta(\hat{F}_m(T)).$$

Let $Y_t = (X_t, X_{t+1}, \ldots, X_{t+m})$, $t = 1, 2, \ldots, T - m + 1$ and let $y \in \Re^m$. We assume that

$$I(y, F(m)) = \lim_{\varepsilon \downarrow 0} \frac{\theta\big((1 - \varepsilon)F(m) + \varepsilon\delta_y\big) - \theta\big(F(m)\big)}{\varepsilon}$$

exists for all $y \in \Re^m$. The function $I(y, F(m))$ is known as the influence function. It is assumed that $E[I(Y_1, F(m))] = 0$. Let

$$\sigma^2 = \text{Var}[I(Y_1, F(m))] + 2 \sum_{s=1}^{\infty} \text{Cov}[I(Y_1, F(m)), I(Y_{s+1}, F(m))].$$

We assume that
$$\sqrt{T}(\hat{\theta} - \theta) \xrightarrow{\mathscr{L}} N(0, \sigma^2).$$

The above is valid for a large class of estimators such that $\hat{\theta} - \theta = (1/T) \sum_{t=1}^{T} I[Y_t, F(m)] + R(T)$, where $\sqrt{T}R(T) \to 0$ in probability.

The block-based bootstrap is defined as follows. Assume that the sample size T can be written as $T = KL$. Let $T^* = T - m + 1$. Let $S_1, S_2, \ldots, S_K$ be i.i.d. random variables with the common distribution as the Uniform distribution on $\{0, 1, \ldots, T - L\}$. The bootstrap version of $\hat{F}_m(T)$ is defined by

$$F_m^*(T) = T^{-1} \sum_{k=1}^{K} \sum_{t=S_k+1}^{S_k+L} \delta_{Y_t}$$

and the bootstrap statistic is defined by

$$\hat{\theta}^* = \theta(F_m^*(T)).$$

In Politis and Romano (1992), such bootstrap is known as blocks of blocks bootstrap and the integer m is allowed to increase with the sample size to include more general estimators. An alternative way of describing the block-based bootstrap is as follows. Let
$$(Y_1^*, \ldots, Y_L^*), (Y_{L+1}^*, \ldots, Y_{2L}^*), \ldots, (Y_{T-L+1}^*, \ldots, Y_T^*)$$

be a SRSWR of size K from the bootstrap population

$$(Y_1, Y_2, \ldots, Y_L), (Y_2, Y_3, \ldots, Y_{L+1}), \ldots, (Y_{T-L+1}, Y_{T-L+2}, \ldots, Y_T)$$

of $T - L + 1$ blocks.

Then, the bootstrap estimator of variance is given by

$$\sigma_B^2 = \text{Var}^*(\hat{\theta}^*).$$

In practice, the above estimator as well the bootstrap estimator of the sampling distribution of $\hat{\theta}$ is computed with the help of simulations. For $m = 1$, if we take $\theta = \int x dF(x) = \mu$, we have $I(x, F(1)) = x - \mu$, and $\hat{\mu} \bar{X}$, the sample mean. In this case, the bootstrap estimator of $\text{Var}(\bar{X})$ can be computed without any simulations. We note that

$$E^*(\bar{Y}^*) = \frac{1}{L(T-L+1)} \sum_{t=0}^{T-L} \sum_{\ell}^{L} X_{t+\ell} \neq \bar{X}$$

$$\text{Var}^*(\bar{Y}^*) = \frac{1}{K(T-L+1)} \sum_{i=0}^{T-L} (\bar{Y}_i - E^*(\bar{Y}^*))^2 \tag{6.1}$$

Properties of block-based bootstrap

The following theorem gives consistency of the bootstrap estimator of variance.

Theorem 6.6.1 (Künsch 1989) *Suppose that $\hat{\theta} = \bar{X}$, the sample mean. We assume that the stationary process $\{X_t, t \geq 1\}$ is strong mixing with*

(i) $E(|X_1|^{8+\delta}) < \infty$ and $\sum_t t^2 [\alpha(t)]^{6/(6+\delta)} < \infty$ for some $\delta > 0$.
(ii) $L(T) = o(T)$ $L(T) \to \infty$.
 Then, the bootstrap estimator $T\sigma_B^2 \to \sigma^2$ in probability. It also converges to σ^2 in the quadratic mean .

Under additional assumptions, Künsch (1989) further proves that the block-based bootstrap estimates the sampling distribution of $\sqrt{T}(\bar{X} - \mu)$ consistently, in the sense that

$$\sup_x \left| P^*\left[\sqrt{T}(\bar{X}^* - \bar{X}) \leq x\right] - P\left[\sqrt{T}(\bar{X} - \mu) \leq x\right] \right| \to 0 \quad \text{a.s.}$$

In block-based bootstrap, it is important to define the bootstrap pivotal carefully. It is *incorrect* to use $\sqrt{T}(\theta^* - \hat{\theta})$, the correct centering is given by $\sqrt{T}(\theta^* - \tilde{\theta})$, where $\tilde{\theta} = E^*(\hat{\theta}^*)$. In general, $\tilde{\theta} \neq \hat{\theta}$, see (6.1).

Bootstrapping empirical process

We now consider block-based bootstrap to approximate distribution of the empirical process of m-dimensional strong mixing sequences. Let F_T^* be as defined earlier and the bootstrap empirical process is defined by

$$G_T^*(y) = \sqrt{T}\left[F_T^*(y) - E^*\left(F_T^*(y)\right)\right], \quad y \in R^m.$$

Theorem 6.6.2 (Bühlmann 1994) *Under the assumptions and notation of Theorem 1.4.1, we further assume that the block size L is such that $L = O(T^{(1/2)-\varepsilon})$ for some*

ε, $0 < \varepsilon < 1/2$. *Then,*

$$G_T^* \to G \ weakly, \quad a.s.$$

The above result for the one-dimensional case has been proved in Naik-Nimbalkar and Rajarshi (1994). In the papers by Naik-Nimbalkar and Rajarshi (1994) and Bühlmann (1994), under less restrictive conditions, the above convergence has been shown to hold in probability. For an exhaustive account of bootstrapping empirical process from stationary sequences, we refer to Radulović (2002).

With the help of the above theorem, one can prove that, if the functional θ satisfies Hadamard differentiability (cf. Fernholz (1983)), the bootstrap gives asymptotically valid approximation to the sampling distribution of $\sqrt{T}(\hat{\theta}_T - \theta)$, see Radulović (2002). As remarked earlier, for studying the sampling distribution of a statistic such as the sample median, block-based bootstrap is very handy, as an analytical study is quite difficult.

Lahiri (2003) establishes consistency of the block-based bootstrap estimator of the sampling distribution of a Fréchet differentiable statistics . His result is as follows.

Theorem 6.6.3 *(Theorem 4.4 of Lahiri 2003) Let $T_0 = T - m + 1$. Under the assumptions of Theorem 1.4.2, we further assume that $L \to \infty$ such that $L/\sqrt{T} \to 0$. Then,*

$$\sup_z \left| P^* \left[\sqrt{T_0} \left(\hat{\theta}^* - \tilde{\theta} \right) \le z \right] - P \left[\sqrt{T} \left(\hat{\theta} - \theta \right) \le z \right] \right| \to 0, \ in \ probability.$$

Hence, in view of the above theorem, the block-based bootstrap gives a consistent estimator of the sampling distribution of $\hat{\theta}$.

Consistency of bootstrap estimator of the sampling distribution of a centralized statistic viz., $\sqrt{T}(\hat{\theta} - \theta)$ has been established in a large number of cases. However, consistency of bootstrap estimator of the variance σ^2 of the asymptotic normal distribution of $\sqrt{T}(\hat{\theta}_T - \theta)$ has been proved for a number of classes of statistics. For a very broad class of estimators, we can write $(\hat{\theta} - \theta) = \sum_{t=1}^{T} \psi(X_t, \theta) + R(T)$. Asymptotic normality is established (for the sample as well as for a bootstrap sample) by showing that $\sqrt{T} R(T) \to 0$ in probability. However, for consistency of bootstrap estimator of the variance, one needs to show that $T E[(R(T))^2] \to 0$ and that $P^*[T E[(R^*(T))^2] \to 0] \to 1$. These convergences appear to be main sources of difficulty. Parr (1985) proposes that the standard deviation σ be estimated as follows. Suppose one has proved that

$$\sup_x \left| P^* \left[\sqrt{T} \left(\hat{\theta}^* - \tilde{\theta} \right) \le x \right] - P \left[\sqrt{T} \left(\hat{\theta} - \theta \right) \le x \right] \right| \to 0 \ \ in \ probability/a.s.,$$

where $\tilde{\theta}$ is an appropriate statistic (which need not be the same as $\hat{\theta}$). Then, the interquartile range of $\sqrt{T}(\hat{\theta}^* - \tilde{\theta})$ (as estimated from a bootstrap array) can be taken as a consistent estimator of σ. This suggestion is based on the fact that interquartile range, unlike the standard deviation, is a continuous function on the space of

absolutely continuous distribution functions, provided each p.d.f. is positive for all
real x.

Second order properties of block-based bootstrap

In the i.i.d. case, as remarked in Sect. 6.1, Efron's bootstrap procedure gives a better
approximation to the sampling distribution of an approximate pivotal than the one
given by the CLT. In the recent years, this result has been extended to block-based
bootstrap in important papers by Götze and Künsch (1996) and Lahiri (1991).

Definition 6.6.1 Smooth function model. We assume that the state-space is $\Re^p$. Let
$\mu = (\mu_1, \mu_2, \ldots, \mu_p)' = E(X_1)$ and let a parameter θ be defined by a function
$H(\mu)$, where $H : \Re^p \to \Re$. Let $\hat{\theta} = H(\overline{X})$. Suppose that H is differentiable in a
neighborhood of μ given by $N = \{x | x \in \Re^p, \| x - \mu \| < \eta\}$ for some $\eta > 0$. Let
$DH(\mu) = (\partial H/\partial \mu_1, \partial H/\partial \mu_2, \ldots, \partial H/\partial \mu_p)'$ be the $p \times 1$ gradient vector of H.
We assume that $DH(\mu)$ is a non-zero vector. The function θ is then said to satisfy
the smooth function model.

A large number of estimators satisfy smooth function model. These include
moment estimators , LSEs , sample auto-correlations and partial auto-correlations
among others. Under appropriate assumptions on rate of decay of mixing co-efficients
(Theorem 1.3.4), we have

$$\sqrt{T}(\overline{X} - \mu) \overset{\mathscr{D}}{\to} N_p(0, \Sigma),$$

and further

$$\sqrt{T}\left(H(\overline{X}) - H(\mu)\right) \overset{\mathscr{D}}{\to} N\left(0, [DH(\mu)]'\Sigma[DH(\mu)]\right).$$

Let

$$\sigma^2 = [DH(\mu)]'\left(\sum_{t=-\infty}^{\infty} E\left[(X_1 - \mu)(X_t - \mu)'\right]\right)[DH(\mu)]$$

$$\sigma_T^2 = [DH(\mu)]'\left(\sum_{t=-T}^{T}\left(1 - \frac{t}{T}\right) E\left[(X_1 - \mu)(X_t - \mu)'\right]\right)[DH(\mu)].$$

It follows that $\sigma_T^2 \to \sigma^2$ as $T \to \infty$. We estimate σ_T^2 (and σ^2) by

$$\hat{\sigma}_T^2 = [DH(\overline{X})]'\left(\frac{1}{T}\sum_{t=1}^{T}(X_t - \overline{X})(X_t - \overline{X})' + 2\sum_{\ell=1}^{L}\sum_{t=1}^{T-\ell}(X_t - \overline{X})(X_{t+\ell} - \overline{X})'\right)$$
$$[DH(\overline{X})].$$

The asymptotic pivotal is given by

$$U_{\text{Student}} = \frac{\sqrt{T}\left(H(\overline{X}) - H(\mu)\right)}{\hat{\sigma}_T}.$$

We implement the block-based bootstrap as described above. Let $\overline{X}^*$ be a bootstrap sample mean and let $H(\overline{X}^*)$ be the corresponding bootstrap statistic. As remarked earlier, $E^*(\overline{X}^*) \neq \overline{X}$. Let $\tilde{X} = E^*(\overline{X}^*)$. It follows that

$$\tilde{X} = \frac{1}{L(T-L+1)} \sum_{t=0}^{T-L} \sum_{s=1}^{t} X_{t+s}.$$

The difference between $\tilde{X}$ and $\overline{X}$ can be shown to be $O_p(L^{1/2}/T)$, which is larger than the first term in the Edgeworth expansion of $\sqrt{T}(\overline{X}-\mu)$. Therefore, it is essential to consider the bootstrap pivotal as

$$U^*_{\text{Student}} = \frac{\sqrt{T}\left(H(\overline{X}^*) - H(\tilde{X})\right)}{\hat{\sigma}^*_T},$$

where

$$\hat{\sigma}_T^{2*} = \frac{1}{K} \sum_{j=1}^{K} \left[\frac{1}{L^{1/2}} \sum_{i=1}^{L} [DH(\overline{X}^*)]'(X_{i+j} - \overline{X}^*) \right]^2.$$

Theorem 6.6.4 (Götze and Künsch 1996) *We assume that the followings hold.*

A1. $E(X_t) = 0$.
A2. $E \parallel X_t \parallel^{s+\delta} < \infty$ *for some* $s \geq 8$ *and* $\delta > 0$.
A3. *There exists a sequence $\mathcal{D}_k$ of sub-σ fields and a constant $d > 0$ such that for $j, m = 1, 2, \ldots$, with $m > d - 1$, the r.v. X_j can be approximated by a $\sigma\{\mathcal{D}_p| |p - j| < m\}$-measurable random vector $\overline{X}_{j,m}$ with $E(\parallel X_j - \overline{X}_{j-m} \parallel) \leq d^{-1}\exp(-dm)$.*
A4. *There exists a $d > 0$ such that for all $m, j = 1, 2, \ldots, A \in \mathcal{D}_{-\infty,j}, B \in \mathcal{D}_{j,\infty}$, we have $|P(A \cap B) - P(A)P(B)| \leq d^{-1}\exp(-dm)$.*
A5. *Let $Z_t = [DH(\mu)]'(X_t - \mu)$. Then, there exists a $d > 0$ such that for all $m, j = 1, 2, \ldots, 1/d < m < j$ and $\tau \geq d$,*

$$E|E(\exp[i\tau(Z_{j-m} + \cdots + Z_{j+m})])|\mathcal{D}_k, k \neq j| \leq \exp(-d)$$

and

$$\liminf_{T\to\infty} \text{Var}(Z_1 + Z_2 + \cdots + Z_T)/T > 0.$$

A6 . There exists a $d > 0$ such that for all $m, j, p = 1, 2, \ldots$ and $A \in \mathcal{D}_{j-p,j+p}$,

$$E\left|P[A|\mathcal{D}_\ell, \ell \neq j] - P[A|\mathcal{D}_\ell, 0 < |\ell - j| \leq m + p]\right| \leq d^{-1}\exp(-dm).$$

A7. *The function* $H : \Re^p \to \Re$ *is thrice differentiable. The vector* $DH(\mu) \neq 0$. *Further, there exist constants* $C_1, C_2 > 0$ *such that* $\| D_3 H(x) \| < C_1 \left(1 + \| x \|^{C_2}\right)$ *for every* $x \in \Re^p$.

Let the block size L *satisfy the conditions that* $L < T^{1/3}$ *and* $\ln T = o(L)$. *Further,* $(\ln T)^M \leq L \leq T^{1/3}$ *for a large enough* M *and the conditions A1–A4 hold with* s *replaced by* qs, $q \geq 3$ *and* $s \geq 8$. *Then,*

$$\sup_x \left| P^*[U^*_{\text{Student}} \leq x] - P[U_{\text{Student}} \leq x] \right| = O_P(T^{(-3/4)+\varepsilon}),$$

where $\varepsilon > 2/s$.

If all the moments of X_1 are finite,

$$\sup_x \left| P^*[U^*_{\text{Student}} \leq x] - P[U_{\text{Student}} \leq x] \right| = O_P(T^{(-3/4)+\varepsilon})$$

for all $\varepsilon > 0$. Since the normal approximation is $O(T^{-1/2})$, it follows that bootstrap approximation gives a better approximation to the sampling distribution of U_{Student}.

The assumptions A3–A6 of the above theorem hold good for Mixing sequences discussed in Sect. 1.3. For verification of these assumptions for various processes including those in Sect. 1.3, we refer to Götze and Hipp (1983); Bose (1988); Götze and Künsch (1996), and Lahiri (2003), Sect. 6.3

Lastly, it needs to be mentioned that Lahiri (1991, 2003). Section. 6.3 proposes a different set of assumptions and norming for the Studentization and proves the second order correctness of the block-based bootstrap. In Lahiri's modification, we use the bootstrap pivotal, given by

$$U^*_1 = \frac{\sqrt{T}\left(H(\overline{X}^*) - H(\tilde{X})\right)}{\hat{\sigma}_T},$$

where $\hat{\sigma}_T$ is a consistent estimator based on the sample and has been used in defining the sample pivotal U_{Student}. In other words, the norming is the same for all bootstrap samples. Lahiri (1991) shows that

$$\sup_x \left| P^*[U^*_1 \leq x] - P[U_{\text{Student}} \leq x] \right| = o_P(T^{-1/2}).$$

This proves that the bootstrap gives a better approximation than the CLT approximation.

Choosing an optimal block size

A natural question is regarding the choice of L which improves the performance of the variance estimator and that of the confidence interval. Hall et al. (1995) discuss sample-based choice of the optimal block size L for Künsch's Block-based bootstrap. The Chap. 7 of Lahiri (2003) gives a thorough discussion of the optimal block size. We outline an empirical method due to Hall et al. (1995) to choose the block length

L for Künsch's bootstrap. Let us suppose that we are interested in estimating the sampling distribution of a Studentized asymptotic pivotal $U_T = (\hat{\theta} - \theta)/\hat{\sigma}_T$, where $\hat{\sigma}_T^2$ is a consistent estimator of σ^2, the variance of the asymptotic normal distribution of $\hat{\theta}_T$. It is assumed that θ satisfies the smooth function model. Let

$$F(x) = P[U_T \le x].$$

Let L be an initial choice of the block length. A primary bootstrap estimator of $F(x)$ is given by

$$\hat{F}(x) = P^*[U_T^* \le x],$$

where $U_T^* = (\hat{\theta}^* - \hat{\theta})/\hat{\sigma}_T^*$. Hall et al. (1995), under the assumptions of Götze and Künsch (1996) described above, show that, for a large T, the MSE of $\hat{F}(x)$ as an estimator of $F(x)$ is given by

$$E\left[\hat{F}(x) - F(x)\right]^2 \approx \frac{1}{T}\left(\frac{C_1}{L^2} + \frac{C_2 L^2}{T}\right),$$

where C_1 and C_2 are functions of large sample moments of the sample mean vector and derivatives of the function of the mean vector. The above quantity is obviously minimized by taking $L \propto T^{1/4}$.

Hall et al. (1995) proceed to suggest the following empirical method to estimate the optimal length L^* of the block size. We regard each block of size L as a sample itself and select $L' < L$ as the block size for this sample. Then, we have $T - L + 1$ estimators of $F(x)$, which we denote by $\hat{F}_s(x)$, $s = 1, 2, \dots, T - L + 1$. We select L_1^* as that value of L' which minimizes $\sum_s[\hat{F}_s(x) - \hat{F}(x)]^2$. The optimal block size for the sample of size T that we have, is then given by $(L/L')^{1/4} L_1^*$.

Hall et al. (1995) also discuss estimation of bias and variance of a statistic and estimation of the sampling distribution of $|U_T|$. For the first two, the optimal block size is of the order of $T^{1/3}$, whereas for the last one, it is of the order of $T^{1/5}$. In each case, an empirical rule to compute optimal block size is similar to the one outlined above. Hall et al. (1995) estimate the optimal L by simulation studies of a model and show that distribution of the empirically obtained optimal block-lengths, as discussed earlier, have a mode at the optimal value.

It has been pointed out in the literature that the block-based bootstrap has some shortcomings. In a bootstrap sample, the first and the last L observations occur with less frequency than the other observations. Further, a bootstrap sample does not form a stationary sequence. Politis and Romano suggest circular and stationary bootstrap to deal with these two problems. In the circular bootstrap procedure (Politis and Romano 1992), blocks are formed by wrapping i.e., attaching the last observations with the first observations. For example, with $T = 9$ and $L = 3$, we have 2 more blocks given by $(X_9, X_1, X_2), (X_8, X_9, X_1)$ and in addition to earlier 7 blocks given by $\{(X_1, X_2, X_3), \dots, (X_7, X_8, X_9)\}$. We select 3 blocks on a SRSWR basis from these blocks, thus the sample size matches with the original sample size 9. In the circular bootstrap, the bootstrap sample mean is conditionally unbiased for the sample mean. In the stationary bootstrap (Politis and Romano 1994), the block lengths are random variables, with the distribution of a block size given by a geometric distribution $p_j = (1 - p)^{j-1} p$, $j \geq 1$, where $L = 1/p$. Thus, in this procedure, p is allowed to converge to 0 such that $pT \to \lambda$, $0 < \lambda < \infty$. The block-based bootstrap due to Künsch and these two bootstraps employ overlapping blocks, one can have non-overlapping blocks also (which will be less in numbers). The bootstrap procedures due to Künsch and the Circular bootstrap outperform the stationary and non-overlapping bootstrap procedures for estimation of bias and variance of a smooth function of the sample mean vector, cf. Lahiri (2003) Chap. 5,

Block-based bootstrap procedures for confidence interval are somewhat handicapped, as the optimal choices for estimation of the variance of a statistic (which is needed in a Studentization procedure) and for estimation of the sampling distribution are of different orders. Davison and Hall (1993) point out that improvement obtained by the block-based bootstrap over the CLT approximation heavily depends upon the way Studentization is carried out and the naive percentile method is as good as the CLT approximation. They remark that the naive bootstrap does not capture the sampling distribution of a Studentized pivotal, since it indirectly assumes that the underlying series is L-dependent. Davison and Hall (1993) suggest a modification to the estimator for the variance of the sample mean. They consider the model given by $X_t - \mu + \sum_j w_j \varepsilon_{t-j}$, where $\{\varepsilon_t\}$ is a sequence of i.i.d. random variables with mean 0 and variance 1 and $\sum_t | t w_t | < \infty$, $\sum_t w_t \neq 0$. The parameter of interest is the stationary mean μ. They show that the Studentized pivotal which uses the modified estimator of the variance (without a bootstrap) leads to an approximation which is superior to the approximation given by the naive bootstrap.

Edge effects and lack of stationarity are some of the difficulties which have been solved to some extent. Bühlmann (2002) brings out limitations of the block-based bootstrap. He points out that performance of block-based bootstrap is not satisfactory for categorical time series and procedures such as AR-seive bootstrap described in Bühlmann (2002), outperform block-based bootstrap.

Results of Davison and Hall (1993) and Bühlmann (2002) suggest that there are some limitations to the block-based bootstrap and in specific cases, there exist alternative techniques which are superior to the block-based bootstrap. Nevertheless, as pointed out earlier, since a theoretical analysis in stochastic models can be quite formidable, a block-based bootstrap procedures without any assumptions (apart from that of stationarity and smoothness conditions of a statistic) are quite useful in practice. They do not need any complicated theoretical computations and we easily get variance estimators and the estimators of the sampling distribution which are typically at least as good as the classical CLT approximation.

6.7 Other Block-Based Sample Reuse Methods for Stationary Observations

Subseries Technique (Carlstein 1986)

Let $K = T/L$. There are K non-overlapping blocks of size L. Carlstein (1986) proposes the variance of the K estimators of θ obtained from these blocks as the estimator of variance of $\hat{\theta}$.

A Direct Estimator

We now explicitly indicate the parameters in the functional form of the Influence function by writing $IF(y, \eta)$ instead of $IF(y, F(m))$. Considering the Influence function form of the variance of asymptotic normal distribution and assuming that the parameters η have been estimated by a consistent estimator $\hat{\eta}$, define an estimator of σ^2 as

$$
\sigma_{\mathrm{Infl}}^2 = \sum_{k=L-1}^{L+1} w(T, k) \sum_{t=1}^{T-|k|} IF(Y_t, \hat{\eta}) IF(Y_t + |k|, \hat{\eta}).
$$

It is assumed that $w(T, k) \to 1$ (for fixed k) as $T \to \infty$ and that $L \to \infty, L/T \to 0$. Smoothness conditions on the influence function are required to show consistency of σ_{Infl}^2. General results are known in those cases where the estimator agrees with the block-based jackknife, which we discuss now.

Block-based Jackknife(Künsch 1989)

Consider the weights $w_T(t)$ for $t = 1, 2, \ldots$ such that $0 \le w_T(t) \le 1$ and $w_T(t) > 0$ if and only if $1 \le t \le L$. Let $\| w_T \| = \sum w_T(t)$. Define distribution functions

$$
F_m(T)^{(-j)} = \frac{1}{T - \| w_T \|} \sum_{t=1}^{T} (1 - w_T(t - j)) \delta_{Y_t},
$$

where, as before, $Y_t = (X_t, X_{t+1}, \ldots, X_{t+m})$. Let

$$
\theta^{(j)} = \theta[F_m(T)^{(-j)}], \quad j = 0, 1, \ldots, T - \ell.
$$

The notation $(-j)$ denotes the property that the j-th block has been deleted while computing an estimator of the d.f. Then, the Jackknife estimator of the variance of $\hat{\theta}$ is defined by

$$
\sigma_J^2 = \frac{1}{(T - \| w_T \|)^2} \frac{1}{T(T - L + 1) W(2)} \sum_{j=0}^{T-L} \left(\theta^{(j)} - \tilde{\theta} \right)^2,
$$

where $W(2) = \sum_{t=1}^{L} [w_T(t)]^2$ and $\tilde{\theta} = (T - L + 1)^{-1} \sum_{j=0}^{T-L} \theta^{(j)}$. We note that $T\sigma_J^2$ estimates the variance of the asymptotic normal distribution of $\sqrt{T}(\hat{\theta} - \theta)$.

The weight functions are of the form $w_T(t) = h((t - 1/2)/L)$, $1 \le t \le L$, where the function h is symmetric about $1/2$ and increasing on $(0, 1/2)$. The simplest choice is h to be indicator function of the interval $(0, 1)$ and it indicates deletion of a block. The block length L satisfies the condition that $L \to \infty$ and $L/T \to 0$.

Künsch (1989) shows that the block-based jackknife procedure gives a consistent estimator of the asymptotic variance for (i) estimators which are smooth functions of the sample mean vector of functions (ii) von Mises statistics with a symmetric kernel (iii) estimating function estimators satisfying certain regularity conditions for observations from ARMA models (Lemma 4.1, 4.2, and 4.3, respectively of his paper).

Sample reuse method(Hall and Jing 1996)
We now describe a procedure which is based on blocks of observations, however, it does not obtain repeated samples from the given set of observations. We suppose that we are interested in estimating the sampling distribution of a Studentized asymptotic pivotal $U_T = \sqrt{T}(\hat{\theta} - \theta)/\hat{\sigma}_T$, where $\hat{\sigma}_T^2$ is a consistent estimator of σ^2, the variance of the asymptotic normal distribution of $\sqrt{T}(\hat{\theta} - \theta)$. Let

$$F(x) = P[U_T \le x].$$

Let $T(1) = T - L + 1$ be the number of (overlapping) blocks, each of length L. Let $\hat{\theta}_s$ and $\hat{\sigma}_s$ be the corresponding estimators of θ and σ respectively, based on the sth block. A primary estimator of $F(x)$ is given by

$$\hat{F}(x) = \frac{1}{T(1)} \sum_s I\left[\frac{\sqrt{L}(\hat{\theta}_s - \hat{\theta})}{\hat{\sigma}_{(s)}} \le x \right].$$

The above estimator needs to be corrected for the fact that the sample sizes do not match. Hall and Jing (1996) suggest Richardson extrapolation procedure to deal with this difference. It is assumed that the pivotal $\sqrt{T}(\hat{\theta} - \theta)/\hat{\sigma}_T$ satisfies conditions of Götze and Hipp (1983) or Götze and Künsch (1996) so that it admits a first order Edgeworth expansion. Then, the modified estimator of $F(x)$ is given by

$$\tilde{F}(x) = \left(1 - (L/T)^{1/2}\right)\Phi(x) + (L/T)^{1/2}\hat{F}(x),$$

where $\Phi(x)$ is the distribution function of a $N(0, 1)$ random variable. Hall and Jing (1996) show that the MSE of $\tilde{F}(x)$ as an estimator of $F(x)$ is minimum when $L = T^{1/3}$ and the minimum value is of the order of $T^{-2/3}$. If we are interested in estimating the distribution of $\sqrt{T}|\hat{\theta} - \theta|/\hat{\sigma}_T$, then the optimal value of L is of the order of $T^{1/7}$ and the minimum is of the order $T^{-8/7}$. They also describe empirical procedures for selecting L.

6.8 Resampling Based on Estimating Functions

Bootstrap and other resampling procedures have been extended by several researchers to estimating functions rather than observations. Bootstrapping estimating functions was discussed by Hu and Zidec (1995) in the case of robust estimators in a regression model. Hu and Kalbfleisch (2000) discuss general results and give several applications. As pointed out by Hu and Zidec (1995), there are two distinct advantages of obtaining confidence interval directly from an estimating function. First, an estimator need not be computed for each bootstrap sample. Thus, the bootstrap is less time-consuming. Second, normal approximation for a Studentized estimating function is often better than normal approximation to the Studentized estimator obtained by solving the corresponding equation.

Delete 1 jackknife for uncorrelated estimating functions
Let us suppose that we have T uncorrelated elementary estimating functions H_t, $t = 1, 2, \ldots, T$ satisfying the regularity conditions. Here, instead of deleting an observation, we delete an estimating function H_t at a time and estimate θ from the remaining $T - 1$ estimating functions. Let $\hat{\theta}_{(t)}$ be such an estimator, $t = 1, 2, \ldots, T$ and $\tilde{\theta}$ denote their mean. The jackknife estimator of variance of $\hat{\theta}$ is then defined by

$$\hat{\sigma}^2(J) = \frac{T-1}{T} \sum_t \left(\hat{\theta}_{(t)} - \tilde{\theta} \right)^2$$

Lele (1991a) proves that under certain conditions, the above estimator is a consistent estimator of variance of the asymptotic normal distribution of the estimator obtained by solving the estimating equation $\sum_t H_t = 0$. It may be recalled that elementary estimating functions which form the CLS estimator and the orthogonal estimation functions in stochastic models are uncorrelated and Lele's procedure can be applied to estimators obtained from such procedures.

Block-based bootstrap for estimating functions
In estimating function bootstrap in the i.i.d. case, instead of resampling from a set of observations, we obtain a SRSWR sample from $\{H_1(\hat{\theta}), H_2(\hat{\theta}), \ldots, H_T(\hat{\theta})\}$, where $H_t(\theta)$ is an elementary estimating function based on the observation X_t alone and $\hat{\theta}$ is a consistent solution of $\sum H_t = 0$.. Here, this procedure is extended to the block-based bootstrap for stationary observations.
Let $Y_t = (X_t, X_{t+1}, \ldots, X_{t+m})$, $t = 1, 2, \ldots, T - m + 1$, as before. We now assume that the elementary function H_t is a function of Y_t and θ (the parameter of interest) only. Define

$$U_t(\theta) = H_t(Y_t, \theta), \quad t = 1, 2, \ldots, T - m + 1, \quad \overline{U}_T = (T - m + 1)^{-1} \sum U_t.$$

Let

$$V_T^2(\theta) = \frac{1}{T - m + 1} \sum_t U_t^2 + 2 \sum_\ell \frac{1}{T - \ell} \sum_t U_t U_{t+\ell}.$$

We wish to estimate the sampling distribution of the Studentized estimating function $G_T(\theta)$ defined as

$$G_T(\theta) = \frac{1}{\sqrt{T}} \frac{\overline{U}_T}{V_T(\theta)}.$$

Under appropriate mixing conditions, moment conditions, and conditions on the block length L (Theorem 1.3.4), it can be shown that (i) $\frac{\overline{U}_T}{\sqrt{T}}$ converges to $N(0, V^2)$ in distribution and that (ii) $V_T(\theta)$ converges to V a.s./in probability. This follows easily from Künsch (1989), since $\overline{U}_T$ is a mean of consecutive observations from a stationary and α-mixing random variables, (cf. Theorem 1.3.4).

In EF bootstrap, we regard $U_t(\hat{\theta})$'s as stationary observations and carry out block-based bootstrap procedure. For the sake of convenience, we write

$$U_t = U_t(\hat{\theta}), \quad t = 1, 2, \ldots, T - m + 1,$$

$$\overline{U}_t = \frac{1}{L} \sum_{s=t}^{t+L} U_s, \quad t = 1, 2, \ldots, T - m + 1.$$

Let $\overline{U}_1^*, \overline{U}_2^*, \ldots, \overline{U}_{T(1)}^*$ be a SRSWR of size $T(1)$ from $\overline{U}_1, \overline{U}_2, \ldots, \overline{U}_{T(1)}$. Let

$$H_T^* = \frac{1}{T(1)} \sum_{t=1}^{T(1)} \overline{U}_t^*.$$

It may be pointed out that, in general, $E^*(H_T^*) \neq 0$, i.e., the estimating function H_T^* is not a conditionally unbiased estimating function. Let $\tilde{H}_1 = E^*(H_T^*)$. Let $\tilde{U}_T = E^*(\overline{U}_1^*)$. Let

$$G_T^*(\hat{\theta}) = \frac{\sqrt{T - m + 1}(H_T^* - \tilde{H}_T)}{V^*},$$

where

$$(V^*)^2 = \frac{1}{T - m + 1} \sum_t \left(\overline{U}_t^* - \tilde{U}_T\right)^2.$$

Then, the bootstrap approximation to the distribution of $G_T(\theta)$ is given by

$$\hat{P}[G_T(\theta) \leq x] = P^*\left[G_T^*(\hat{\theta}) \leq x\right],$$

which, in practice, is approximated by simulations.

A generalized bootstrap for estimating equations

Recently, Chatterjee and Bose (2005) have introduced a generalized bootstrap procedure which is based on estimating equations. This procedure is different than those by Hu and Kalbfleisch (2000) and Lele (1991b). Let $\{H_t, t \geq 1\}$ be orthogonal estimating functions i.e., $E[H_t|X_0, X_1, \ldots, X_{t-1}] = 0$. Let $\{W_t, t = 1, 2, \ldots T\}$ be a collection of random variables (which depend on T). The random variables $\{W_t, \geq 1\}$ are independent of observations and independently distributed of $\{H_t, t \geq 1\}$. These are known as bootstrap weights. Earlier work wherein such bootstraps were studied are Freedman and Peters (1984) and Rao and Zhao (1992). These weights satisfy the following conditions.

A1. For each T, $\{W_t, t = 1, 2, \ldots T\}$ are exchangeable random variables. $E[W_t] = 1$ for all t.

A2. Let $\text{Var}(W_t) = \sigma_t^2$. Then, $E(W_1 W_2) = O(1/T)$.

A3. $E(W_1^2 W_2^2) \to 1$ and $E(W_1^4) < \infty$.

The above conditions are satisfied by Efron's bootstrap, jackknife, and a number of resampling plans. For example, Efron's bootstrap is obtained when $\{W_t, t = 1, 2, \ldots T\}$ has a multinomial distribution with parameters T and the cell probabilities $(1/T, 1/T, \ldots, 1/T)$. We confine ourselves to the case when θ is a scalar parameter. Results of Chatterjee and Bose (2005) are more general and are applicable to a non-stationary processes.

Let us assume that the underlying sequence is stationary.

A4. Each $H_t(\theta)$ admits a first order Taylor series expansion in a neighborhood of the true parameter.

A5. $E[\sum_{t=1}^{T} H_t^2] \to \infty$. This is trivially satisfied in the stationary case, since $0 < E(H_t^2) < \infty$.

A6. Let us write the Taylor series expansion as $H_t = H_t(\theta_0) + (\theta - \theta_0)H_t'(\theta_0) + (1/2)(\theta - \theta_0)^2 H_t''(\theta^*)$, where θ_0 is the true value and θ^* is the intermediate point. Then, $\sup_{|\theta - \theta_0| < \delta} H''(\theta) < M(\theta_0)$, where $E[M(\theta_0)^2] < \infty$.

Chatterjee and Bose (2005) (cf. their Theorems 3.1 and 3.2) show that the equation $\sum H_t = 0$ admits a consistent and asymptotically normal solution $\hat{\theta}$ and the weighted bootstrap approximation to the distribution of the appropriately Studentized $\hat{\theta}$ is asymptotically valid. Their results are more general and cover non-stationary processes also.

References

Allen, M., Datta, S.: A note on bootstrapping M-estimators in ARMA models. J. Time Ser. Anal. **20**, 365–379 (1999)

Athreya, K.B., Fuh, C.D.: Bootstrapping Markov chains : countable case. J. Stat. Plan. Inf. **33**, 311–331 (1992a)

Athreya, K.B., Fuh, C.D.: Bootstrapping Markov chains : countable case. In: LePage, R., Billard, L. (eds.) Exploring the Limits of Bootstrap, pp. 49–60. Wiley, New York (1992b)

Basawa, I.V., Mallik, A.W., McCormick, W.P., Taylor, R.L.: Asymptotic bootstrap validity for finite Markov chains. Commun. Stat. Theor. Meth. **19**, 1493–1510 (1990)

Bickel, P.J., Bühlmann, P.: A new mixing notion and functional central limit theorems for a sieve bootstrap in time series. Bernoulli **5**, 413–446 (1999)

Bose, A.: Edgeworth correction by bootstrap in autoregression. Ann. Stat. **16**, 1709–1722 (1988)

Bose, A.: Bootstrap in moving average models Ann. Inst. Stat. Math. **42**, 753–768 (1990)

Brockwell, P.J., Davis, R.A.: Time Series : Theory and Methods. Springer, New York (1987)

Bühlmann, P.: Blockwise bootstrapped empirical process for stationary sequences. Ann. Stat. **22**, 995–1012 (1994)

Bühlmann, P.: Bootstraps for time series. Stat. Sci. **17**, 52–72 (2002)

Carlstein, E.: The use of subseries values for estimating the variance of a general statistic from a stationary sequence. Ann. Stat. **14**, 1172–1179 (1986)

Chatterjee, S., Bose, A.: Generalized bootstrap for estimating equations. Ann. Stat. **33**, 414–436 (2005)

Chernick, M.R.: Bootstrap Methods : A Guide for Practitioners and Researchers, 2nd edn. Wiley, New York (2008)

Davison, A.C., Hall, P.: On Studentizing and blocking methods for implementing the bootstrap with dependent data. Aust. J. Stat. **35**, 215–224 (1993)

Davison, A.C., Hinkley, D.V.: Bootstrap Methods and Their Application. Cambridge University Press, Cambridge (1997)

Efron, B., Tibshirani, R.: An Introduction to the Bootstrap. Chapman and Hall, London (1993)

Fernholz, L.T.: Von Mises Calculus for Statistical Functionals. Lecture Notes in Statistics 19. Springer, New York (1983)

Freedman, D.A., Peters, S.C.: Bootstrapping a regression equation : some empirical results. J. Amer. Stat. Assoc. **79**, 97–106 (1984)

Götze, F., Hipp, C.: Asymptotic expansions for sums of weakly dependent random variables. Z. Wahr. verw. Geb. **64**, 211–240 (1983)

Götze, F., Künsch, H.R.: Second-order correctness of the block-wise bootstrap for stationary observations. Ann. Stat. **24**, 1914–1933 (1996)

Hall, P., Jing, B.: On sample reuse methods for dependent data. J. Roy. Stat. Soc.: Ser. B **56**, 727–738 (1996)

Hall, P., Horowitz, J.L., Jing, B.: On blocking rules for the bootstrap with dependent data. Biometrika **82**, 561–574 (1995)

Hu, F.: Efficiency and robustness of a resampling M-estimator in the linear model. J. Multivar. Anal. **78**, 252–271 (2001)

Hu, F., Kalbfleisch, J.D.: The estimating function bootstrap (with discussion). Can. J. Stat. **28**, 449–499 (2000)

Hu, F., Zidec, J.V.: A bootstrap based on the estimating equations of the linear model. Biometrika **82**, 263–275 (1995)

Kreiss, J.P., Franke, J.: Bootstrapping stationary auto-regressive moving-average models. J. Time Ser. Anal. **13**, 297–317 (1992)

Kulperger, R.J., Prakasa Rao, B.L.S.: Bootstrapping a finite Markov chain. Sankhyā A **51**, 178–191 (1989)

Künsch, H.R.: The jackknife and the bootstrap for general stationary observations. Ann. Stat. **17**, 1217–1261 (1989)

Lahiri, S.N.: Second order optimality of stationary bootstrap. Stat. Probab. Lett. **11**, 335–341 (1991)

Lahiri, S.N.: Resampling Methods for Dependent Data. Springer, New York (2003)

Lele, S.R.: Jackknifing linear estimating equations: asymptotic theory and applications in stochastic processes. J. Roy. Statist. Soc.: Ser. B **53**, 253–267 (1991a)

Lele, S.R.: Resampling using estimating functions. In: Godambe, V.P. (ed.) Estimating Functions, pp. 295–304. Oxford University Press, Oxford (1991b)

Liu, R.Y., Singh, K.: Moving blocks jackknife and bootstrap capture weak dependence. In: LePage, R., Billard, L. (eds.) Exploring the Limits of Bootstrap, pp. 225–248. Wiley, New York (1992)

Naik-Nimbalkar, U.V., Rajarshi, M.B.: Validity of blockwise bootstrap for empirical processes with stationary observations. Ann. Stat. **22**, 980–994 (1994)

Paparoditis, E., Politis, D.N.: The local bootstrap for Markov processes. J. Stat. Plan. Inf. **108**, 301–328 (2002)

Parr, W.C.: The bootstrap: some large sample theory and connections with robustness. Stat. Probab. Lett. **3**, 97–100 (1985)

Politis, D.N., Romano, J.P.: A circular bootstrap resampling procedure for stationary data. In: LePage, R., Billard, L. (eds.) Exploring the Limits of Bootstrap, pp. 263–270. Wiley, New York (1992)

Politis, D.N., Romano, J.P.: The stationary bootstrap. J. Amer. Stat. Assoc. **89**, 1303–1313 (1994)

Radulović, D.: On the bootstrap and the empirical processes for dependent sequences. In: Dehling, H., Mikosch, T., Sørensen, M. (eds.) Empirical Process Techniques for Dependent Data, pp. 345–364. Birkhäuser, Boston (2002)

Rajarshi, M.B.: Bootstrap in Markov sequences based on estimates of transition density. Ann. Inst. Stat. Math. **42**, 253–268 (1990)

Rao, C.R., Zhao, L.C.: Approximation to the distribution of M-estimates in linear models by randomly weighted bootstrap. Sankhyā A **54**, 323–331 (1992)

Rohan, N., Ramanathan, T.V.: Order section in ARMA models using the focused information criterion. Aust. NZ. J. Stat. **53**, 217–231 (2011)

Serfling, R.J.: Approximation Theorems of Mathematical Statistics. Wiley, New York (1980)

Shao, J., Tu, D.: The Jackknife and Bootstrap. Springer, New York (1995)

Singh, K.: On the asymptotic accuracy of Efron's bootstrap Ann. Stat. **9**, 1187–1195 (1981)

Index

M. B. Rajarshi, *Statistical Inference for Discrete Time Stochastic Processes*,
SpringerBriefs in Statistics, DOI: 10.1007/978-81-322-0763-4,
© The Author(s) 2012